DEEP IMPACTS

DEEP IMPACTS

Sergio Correa

Copyright © 2012 by Sergio Correa.

ISBN:	Softcover		978-1-4691-5606-4
		Ebook			978-1-4691-5607-1

All rights reserved. No part of this book may be reproduced or transmitted in any form or by any means, electronic or mechanical, including photocopying, recording, or by any information storage and retrieval system, without permission in writing from the copyright owner.

To order additional copies of this book, contact:
Xlibris Corporation
1-800-618-969
www.Xlibris.com.au
Orders@Xlibris.com.au
501342

Contents

Comet Lovejoy 11

Is The Theories On Comet Size Wrong? 15

What Are Comets? 18

Edmund Halley 34

The Day The Earth Stood Still 36

Letters By Henry A. Eckstein 43

Letters By Henry A. Eckstein 47

Some Discussion Taken Place At Above Top Secret 52

Moffett Federal Airfieldnasa Ames Reseasrch Center 57

Image Build 59

UFO Sitings By Apollo 11 Astronauts 61

Richard C. Hoagland Contribution to Nasa 66

Climate Change Disclosure 71

Robbing Pluto of its Planethood 76

Thank you all for all your support and love.

Since I was very young I have always had an interest in astronomy and back then in the sixties I used to watch Star Trek and after seeing a movie at the Queens cinema in Gibraltar I would hurry up to the shop and buy lot of Churros which are very popular there and I would watched to boldly go where no man has gone before and share the Churros with my family and friends. But it was not until we were in England in 1969 when I saw my heroes Neil Armstrong, Michael Collins and Edwin E. Aldrin, Jr. That I realise that planetary travel could well become a reality and we could soon visits other planets and stars. Then in Australia back in the early seventy I remember reading the news papers as well as in all the local channels in which the Catholic Church wanted to stop a program from being televised and they also wanted to keep Mr. Erick Von Daniken out of Australia, they used just about everything to discredit him including calling him an Antichrist and at the time I was not sure what all the commotion was about. Until I saw the Chariots of the Gods, in that movie there were questions that made me wonder if what we had been taught in schools were not complete such as those massive ancient structures such as the Egyptian Pyramids which in our schools the teachers used to say that those huge buildings were built with the use of huge manpower using ropes for pulling and lifting copper, and stone tools to cut those stones. However privately our teachers did not agree with the scientific Egyptian orthodox views but they just did as were told to do. This is why Mr. Erich Von Daniken open a whole new world and offer the idea that our world is just but one of many worlds and that there are other worlds out there in the cosmos, and some planets would be more advance than we are. I am no stranger in seen back then strange things in the night skies and I have always been interested ufologie including the paranormal. I have also capture UFO`s on tape and photos as well but I have never been interested in joining any group thus far.

So now that we are here my idea in this book is to make it as easy as I can for the ordinary person man woman and child out there, so its going to be without the jibber jabber that you get out there and we are going to get straight into the particular subject. In fact I will like to dedicate this book to you, so come onboard and journey to the unknown the simple fact that we are not alone.

Acknowledgement

I like to Acknowledge **Mr. Henry A. Eckstein** for his support and I am immensely grateful.

I like to thank also **Mr. Kirk M Rogers** *www.kiroastro.com* for letting me use of his picture.

Image credit **Mr. Henry A. Eckstein** for the Photographs on the cover book.

Credit to **Geo-engineering May 2009 by the Late Mr. Zecharia Sitchin Global Warming**

>Credit: National Aeronautics and Space Administration
>**Title The Moon By Author Mr. Richard C Hoagland**

>Credit: National Aeronautics and Space Administration
>**VI. GUEST INVESTIGATORS AND VISITORS**

>Credit: National Aeronautics and Space Administration
>The Orbiter *Enterprise*

Comet lovejoy

Comet lovejoy was discovered by Australian amateur astronomer Mr. Terry Lovejoy on December 2, last year 2011. According to NASA scientists it was first calculated that the solid body of cometlove joy had a diameter of 100 meters they then change its size to 200 meters and finally the change it size again to 500 meters. Why? Because this tiny little rock has done something which got NASA scientist virtually with their hands down. This huge comet was filmed by a number of space crafts but also including the Solar and Heliospheric Observatory known as SOHO. Since news of this comet went out by the NASA official which capture my attention especially given the many millions of dollars that the American tax payers has put in to NASA. This statement was made by Karl Battams from the Naval Research Lab in Washington DC it read as follows

Quote: It's absolutely astounding," says Karl Battams of the Naval Research Lab in Washington DC. "I did not think the comet's icy core was big enough to survive plunging through the several million degree solar corona for close to an hour, but Comet Lovejoy is still with us." Unquote
http://www.nasa.gov/mission_pages/sunearth/news/comet-lovejoy.html

This picture of our sun and the comparisons of all our planets in our solar system including Pluto was created by Kirk M Rogers is with kind permission of Mr. Kirk M Rogers www.kiroastro.com.

Take a look at the picture above, you will see in this scale the differences in sizes between our Earth as compare to our other planets including our sun. It may pay you to understand that or though our sun is huge compare to our earth and all the other large planets in our solar system, in fact our sun is comparative small and when you placed it next to such suns/stars as Sirius which is very much bigger than our sun followed by Pollux many time larger than our sun, Arcturus which is around 26 times larger than our sun. Once you reach the scale of Arcturus you then reach other massive humongous suns which dwarf our sun to the point that our sun becomes tiny or to put it more bluntly microscopic would be the honest way to describe it when you compare it to such stars as, Rigel, Aldebaran, and the largest of this to date is Antares which is many hundreds of times larger than our sun. Could it be that some of those huge comets that occasionally head towards our sun are cause by those enormous suns such as those mentioned above? Is it conceivable possible that our sun is also orbiting another star causing those mysterious comets to head our way?

Now look how small is our Earth in comparison to our sun, the size of our Earth is so insignificant in size to our sun that you can burly see it. Lets compare the following 120,000 Km divided by Earth diameter at the equator which is 12756 Km = 9.4073377.

In other words 9.40 sizes of our planet would equal the distance of 120,000 Km from the sun surface now take a ruler and measure the Earth on the above picture roughly less than 1 mm but just say that for the sake of our observation based on above picture, two Earths equal 1 mm and you will get a total of 4.7 mm and when you placed a ruler on this picture of the sun you know that any icy object at that sort of depth would be instantly vaporise consume by both the massive temperature and gravity of our sun and it does not take a person with a doctorate degree to understand this simple facts which the scientific community seem unable to comprehend.

Let me put this apophasis to Karl Battams of the Naval Research Lab, If planet Earth were to be put in orbit around our sun at the distant of 120,000 kilometres from the sun surface, what would happened to our Planet? Would our planet survived? What would happened to the Earths magnetosphere? How long would it take for our mighty oceans to vaporise? Earth ice caps: Antarctic 13.5 million square kilometres, Greenland has 1.7 million square kilometres, Canadian artic has 150,000square kilometres need I add more? How long would it take for our poler caps to melt? How long would it take for our planet to vaporise and perhaps plunge into our sun? or Earth could become another planet size comet.

Its is such a weird statement that Karl Battams has made and even today people with Doctorate degrees and PhDs still hold to the absurd views that a little ice comet can survive the massive temperature of the corona. Comet lovejoy was caught on film heading straight towards the sun and reach 120,000 kilometres above the sun surface and instead of plunging into the sun enormous gravity it miraculously sped back from the massive gravity of the sun as if hit by opposing magnetic poles, How is this possible? how is it that the massive gravity of our sun did not pull comet lovejoy into the arms of our sun? Now comet lovejoy has prove again that comets are not dirty little ice and water snowball but more than that it is astonishing how those scientists are not able to see what the average man and woman on the street including children may I add can see for themselves that the dirty snowball theory does not work, never has and there should be review and thrown were all bad science belong, in the rubbish bin.

Also NASA has not provided any answer into how is it possible that such a comet that was heading straight towards the sun and once it reach the suns immense gravity at a mere distance of only 120,000 km from the surface it should had plunged into the sun, any scientist care to explain this bizarre event?

Perhaps we may be missing a piece of a puzzle here, perhaps we would do better at researching alternative theories which NASA scientists may not have considered and their scientists are looking at this problem in a tunnel vision view instead of looking at other alternatives that would help us to solve this mystery and understand our universe.

Is the theories on comet size wrong?

Today's Astronomers alleges that comets are members of our solar system but that they are very small member with an average size of a mile or two across and that comets are very small snow balls. Even to this very day students are being taught this theorise in schools, colleges by Astronomers In fact for you to get your PhD you need to comply with this theory. Two examples of this are, according to scientists comet Halley which according to scientists is the biggest which is about 10 miles across and the smallest called comet Hartley and is estimated to be about a mile across. And for decades scientist such as Dr. David Morrison has declare that comets are nothing more than little dirty snowballs. In fact Nasa took photos of those two celestial comets and none of them look like they are made of ice and water, or look anything like the dirty snowball of which they are continually mentioning. According to NASA comets are celestial objects which are mainly compost of ammonia, methane, carbon dioxide, and water. So scientists have had for this reason determined that a comet is merely just a little snow ball or dust and ice But is that theory sustainable in view of what we know to date? Also are the NASA scientists putting forward the correct sizes of comets?

The Idea that comets were just tiny dirty snow balls was first produced by the Late Dr. Fred Lawrence Whipple, Dr. Fred Lawrence Whipple was an eminent Astronomer, from November 5, 1906 to August 30, 2004 he worked in the Harvard College Observatory. and get this he worked at the, Harvard College Observatory for seventy years in all. Dr. Fred Lawrence Whipple had a very impressive bio. He was born in November of 1906, in Ottawa and was the son of a farmer. Whipple Studied in the University of California in Los Angeles, at the Occidental College. In 1927 Whipple graduated and in 1931 Whipple obtained a PhD in Astronomy at the University of California. Whipple was also involved in mapping the orbit of the newly discovered planet called Pluto. Later in 1931 Dr. Fred Lawrence Whipple joined the Havard College Observatory to study the trajectories of the meteors as well as confirming that meteors originated in our own solar system instead of from some other region of outer space such as interstellar space.

In 1933 Dr. Fred Lawrence Whipple he did two amazing discoveries, he first discovered a comet which had a repeated cycle around our solar system which was named 36P/Whipple later he also discovered an asteroid which was named 1252 Celestia.

Later on Dr. Fred Lawrence Whipple with amateur astronomer Leslie Peltier would go on to discovered other comets which were non periodic comets in other words those other comets did not repeated their cycles.

Amongst Dr. Fred Lawrence Whipple other accomplishments' he was also an inventor. In 1948 Whipple was awarded the Certificate of Merit. His invention was a tin foil device that was used during the last World War to confuse the enemy radar tracking of the Allied aircraft.

Whipple went on to invent the Whipple shield his device invention help to protect spacecraft by vaporizing small particles.

Whipple also asserted that the Comet nucleus were between a few hundred meters to a tenth of a kilometres across and its just compose of ice, and small rocky particles.

To give you an example that science is continually evolving much like our human evolution is. If for example you would have ask Dr. Fred Lawrence Whipple, and pose the question on whether the moon has an atmosphere? or if the moon is dead? The result answer would had been on the latter. You would had been taught that the moon is just a natural satellite of our planet, visible by the reflecting sun light, and having elliptitical orbit. Including you would had been taught that the moon is approximately 221 600 miles = 356 630.63 kilometres and about 252 950 miles = 407 083.565 kilometres at the apogee. With a diameter of 2160 miles = 3 476.18304 kilometres, so on and so on.

However if back in those years you would had theorise that the moon has an Ionosphere and lets assume just for the sake of the argument that you happened to be a Scientist, a Teacher, or Professor most likely your job would had been terminated and you would had been chastised by the scientific community of that time.

It is a fact that those sort of Cover-Up/Harassments are still going on amongst the scientific community for example the one professional that comes to mind is Virginia Steen McIntyre, PhD. Virginian Steen was working for the, United State Geological Survey and her team of experts was send there to date the spear points at Hueyatlaco in Mexico. At first Virginia Steen McIntyre, PhD thought that the aged of those findings were 20,000 years old but later their research team concluded that the age of those artefacts were 250,000 years old. So Virginia Steen McIntyre, PhD data work in Hueyatlaco was suppressed and she could no longer work in her field.

So is it a fact that if you are a Student, scientist, professor, or teacher that you will need to follow the line in order to stay in your profession? Or to gain a PhD in your chosen profession?

That's that give you food for thought?

So coming back to our discussions regarding the question on whether the moon has an Ionosphere the answer is yes. This evidence first came in in 1970 from the soviet probe Lunar 19 and the other Luna 22. In fact who is to say that perhaps one day we will discovered that our moon is not a dead rock orbiting our planet but that just perhaps it may even have a thin layer of atmosphere as well as water, and snow, or even contain precious minerals that we in the future may be able to mine one day.

What are Comets?

The current view are that Comets are made of three basic parts a solid crust called the nucleus, which is surrounded by a coma. This coma can be huge in diameter, it can reach up to 1 million miles = 1 609 344 kilometres miles in diameter. The tail is the next part of a comet which stretches out from the coma when its close enough to the sun. Its generally accepted view that a comet size ranges from a mere 40 to one hundred kilometres, and for the most part comets are believe to be made of ammonia, methane, carbon dioxide, water, as well as the solid crust which is made of rock however from the data collected by DEEP INPACT it has reveal that very little water has been found in comets and so the question arises, if comet has almost no water and ice how is it that its able to generate such a humongous tail, and coma? From the earliest recorded human history comets have brought fear, doom, but comets has also believed to bring good omens as well. in fact the word comet means haired stars which is were comets got its name from. This comets have been referred to as fossils of our planetary system in which our sun together with all our planets and other celestial bodies such as comets, meteors and possible rogue planets orbits our solar system, Then of course there is the theory as stated in the Late Dr. Carl Sagan book called Comet which was co wrote by his loving wife Ann Druyan. In that book they refer to the possibilities of our sun being part of a binary system which is very probably since Astronomers have discovered binary systems in so many stars system and to this day scientists are still discovering more binary star systems. However at this stage I don't think that astronomers have found a star that is accompanying our sun but since science is continually evolving perhaps one day we may discover that we do indeed have a companion star. In reading the book by the late Dr. Carl Sagan and the late Dr. I. S. Shklovskii, the book is called, Intelligent Life In The Universe on page 460 on fig 33-5 Carl Sagan makes the observation that a Sumerian Akkadian cylinder shows indeed a sun together with other circle objects which Carl Sagan believed are celestial objects or planets In this depiction from the Sumerian seal there are a total of nine large spheres and two smaller spheres.

The Astronomical implications from the Sumerian Akkadian cylinder has for our scientific community and our history is plainly obvious to behold. If the Sumerians had knowledge of the solars system and the stars then how did the Sumerians of 6,000 years ago obtain such information's? Is it possible that the Sumerians of 6,000 years ago obtain their knowledge of the stars and planets by a an extraterrestrials civilisation which the Sumerians worship them as Gods?

On the other hand Dr. Mike Heiser who has an M.A and a Ph.D. in the Hebrew Bible and Ancient Semitic Languages from the University of Wisconsin in Madison which he claimed to be an expert in ancient Sumerian texts has a different theory regarding the Sumerian Akkadian cylinder he believe that the large celestial sphere which Dr. Sagan refers to as the sun is not the sun at all and that the other nine planets and the two smaller celestial spheres which Dr. Sagan declares to be planets, are not planets those dots are just depiction of the Pleaides star system. Of course both Dr. Carl Sagan and Dr. Mike Heiser cannot be correct but if you had to choose between them based on your own observations, who would you trust the most?

Could it be that the one of our most distinguished scientist that we ever had Dr. Carl Sagan be wrong in his descriptions of the Sumerian Akkadian cylinder? I was observing the Planetary Data System logo from Nasa and when observing this emblem it reminds me so much of the Sumerian Akkadian cylinder. It has all our nine planets including our moon and a very strange red comet it appears to me as if that logo is just a representation of the Sumerian Akkadian cylinder seal. Why would Nasa make up a logo that resembles those of the Sumerians?

The Oort cloud Theory

It is widely accepted by the mainstream scientific community (but it should be mention that its only a theory that so far no one has ever seen or detected the Oort Cloud) that the Oort cloud myth which scientists alleged to be compost of two regions, the spherical outer region called outer Oort cloud and the other is called the disc shape inner Oort cloud. The Oort cloud is alleged to be at a distance of around 50,000 Astronomical Units.

One unit is equal to the distance from Earth to our Sun. 1 Astronomical Unit = 149 598 000 kilometres

This Oort Cloud was the idea of the Late, Professor, Jan Henrik Oort, he was a Dutch Astronomer in which may I add he also was the pioneer in the Radio Astronomy field.

The Late Professor, Jan Henrik Oort has a very impressive Biography, he was born April 28 1900 and died in November 5 1992.

Later at the early age of just 17 years old Professor, Jan Henrik Oort obtain a degree in astronomy at the Groningen University in the year 1921.

He then spent one year at the Groningen University as an assistant, later on Oort spent two years at the Yale Observatory. He was later given a position at the Leiden Observatory. Oort other interests was radio waves and thus after World War II Oort Began to use the radio antenna left behind by the Germans and started to work in a new field called radio astronomy.

In the ninety fifties he was also responsible for raising much needed money for the Received a number of honours and awards, including in 1942 Bruce Medal of the Astronomical Society of the Pacific, 1946 Gold Medal of the Royal Astronomical Society, 1951 Henry Norris Russel LECTURESHIP OF THE American Astronomical Society, 1960 Gouden Ganzeveer, 1972 Karl Schwarzschild Medal of the Astromische Gesellschaft, and the 1984 Balzan Price for Astrophysics.

There was the Asteroid 1691 Oort named after him as well as the, Oort Cloud, and including the Oort Constant. However with respect to the late Professor, Jan Henrik Oort I find his theory on the Oort cloud does not hold water. Its true that the mainstream scientific community has accepted his theory and they are even teaching the Oort cloud theory as fact. But has Astronomer seen this Oort cloud?

After all Its all well and good to accept someone theories after they have eventually been prove and observed, but has any Astronomer observed the Oort cloud?

So if any scientist claimed that the Oort cloud is a fact but has not seen or observed it then the whole idea of the Oort cloud should be review and investigated further.

I suppose that the whole Idea of the Oort cloud was that he needed to explain the long period comets and entering our planetary zone from such a trajectory that he thought that from a zone which should be between around 2,000 to 100,000 Astronomical Units. Compare those long period comets to that of Halley's Comet which achieves perihelion meaning that Halley's Comet is near to our sun every 77 years plus or minus in one solar orbit.

Even today there are scientist which still disputes that the Oort Cloud is very real and exists, that is why this Oort cloud theory was invented because the orthodox scientists could not explain were all of this comets were coming from.

The most logical conclusion to this is that there are massive amounts of interstellar comets wandering through out space journey in the cosmos by electrical currents and the electrical fields of energy in which those comets travel from one sun to another.

For those of you that support this theory from the orthodox scientific community, or if you are a teacher and are educating your students in astronomy and may I add you are teaching this study of the Oort cloud as if it were a fact I pose to you this question. Have you ever actually seen or observed the Oort Cloud? If not how do you know it is there? So I put it to you again, if you have not observed or seen this Oort cloud, how do you know that this Oort cloud exist at all? Is it possible that all you are doing its just repeating what someone else has taught you without you yourself having done any research into whether this theory is real or not? Is this what science is all about to accept the theory of someone long time dead. The Oort cloud came about from the fact that comets enter the planetary zone on such a trajectory that the Late, Professor, Jan Henrik Oort must have believe that those comets came from an zone between 2000 and 100,000 Astronomical Units for those of you that don't understand what an astronomical unit is and I don't blame you it looks as if we are talking a language that It's not really English, (1 Astronomical Unit = 149 598 000 kilometres).

These are called long-period comets, because the time taken for them to pass through our part of the solar system again is measurable in millions of years. Compare that to Halley's Comet which is one of the smallest comets we ever seen which achieves perihelion every 76 or 77 years or so. So, some people dispute that the Oort Cloud exists well if they claim that they must have either seen or observed it, saying that these comets simply come from interstellar space. Gravitationally, the Sun's influence is vast, spreading out quite a long way towards the nearest star. However that does not explain how comet lovejoy did defied the enormous gravitational field of our sun when it came on December 2, last year 2011 towards our sun at a mere distance of just 120,000 kilometers from the suns surface, and But that doesn't necessarily mean that it has this great, but very dispersed, pool of comets orbiting around it. Perhaps there are just a massie amount of interstellar comets wandering through space which occasionally make their way into the solar system.

So after explaining all of this on another subject, one of the pieces of evidence for a planet X body is that the distribution of these long-period comets is non-random: where one would expect an even spherical distribution, the pattern is not entirely even, and that anomaly has suggested to some very serious astronomers that a massive planet X size object is out there and is heading this way. I believe that is the case, although it seems unlikely to the point of impossibility that such a thing could be observed by SOHO. I would say that this massive object would had to be a very long way away.

For every PhD., there is an equal and opposite PhD.

Gibson's Law

I like to present to you Professor James McCanney, this distinguished scientist is one of the world most brilliant man in the World to today his has a most impressive bio and Its only proper that his credential be mention in here so here we go. Professor James McCanney has taught the following mathematics courses at the University level in addition to Physics, Computer Science and Astronomy; Abstract Algebra, Linear Algebra, Matrix Algebra, Probability and Statistics, Statistics for Computing, Mathematical Logic, Theory of Numbers, Calculus I, II and III, Engineering Math I and II, Advanced Topics in Geometry and Topology. He has worked about half of his lengthy career in private industry. Much of this was accomplished in multi-lingual settings, having worked in the USA, Latin America and with high-level Russian scientists. He has presented his research at international conferences and is a regular presenter at American Geophysical Union meetings. He has also lectured at Los Alamos National Laboratories, the Air-Space/America International Air show and International Electric Propulsion conferences. Understanding his background is important in placing his new book into perspective. The theoretical work you will encounter here is one of the best kept secrets in the Astronomy world. It was developed in the late 1970's while the author was on the faculty of the

Physics and Mathematics Departments of Cornell University, Ithaca, N.Y. My work was in theoretical Celestial Mechanics and Plasma Physics (for the layman, these are the the studies of planetary motion and electrified gases in outer space).

If Professor James McCanney theories are correct it may explain why comet lovejoy

Did not plunge into the sun. One of his theories which he has done much work on has to do with the way a comet may interact is by an electrical manner with our sun and with other suns out there in the cosmos that comets are associated with.

Professor McCanney has since the year 1979 he was studying physics including mathematics, including the physics of the celestial movements, such as suns, planets, moons, comets which I believe it is known as the, celestial mechanics. At this time Professor McCanney was also studying the nature of the plasma physics which I believe has to do with electro magnetic gases that basically spread into outer space.

Please note that back then in the 70`s this sort of topics was not very well known about nd the view from the scientific community regarding this subject would be in the same class as mysticism. So people in reality would not care to know anything about it and so most scientist would not even care to understand it in any shape or form and the few scientist that were working in those fields would not come out openly instead they would just remain inconspicuous for obvious reasons.

Before I continue I just want to add that even now there are people out there that will stop at nothing the hurt Professor McCanney character in the internet and other forums which is a real shame given that we are now in the 21 century but there are still those individuals that still maintain like those of their predecessors, that the world is still flat, okay lets move on.

So what Professor McCanney did was to had included in his research was to include Electrical and Magneticall fields and he was the very first scientist to have experimented on those fields which had to do with the celestial objects such as the suns, planets, moons, and so on. At this particular time Professor McCanney was then taught at the Cornell University as an instructor, which indecently it was also the same university that the late Dr. Carl Sagan studied. It was that at the Cornell University data from the voyager spacecraft including other spacecrafts as well would be send there. So all the data from the NASA space program was deposit there to have them analyse. It was in those data send to the Cornell University that Professor McCanney

The sort of things that he was working on his theoretical research that he was doing at the time. It is interesting to note that back then astronomy was very much in its infancy and the scientific community then would only accept the gravitational forces in space and not much else, this is what I mean when I say that science is always involving, I don't mean it as some sort of insult but as a reality and it's a fact of life, that when ever you make a new discovery or something else that its not accepted by the mainstream, people are going to be very sceptical and there will be those who would take offence especially when you prove them wrong. So as you can see Professor McCanney started to know and to understand were those electric fields were coming from in the cosmos so Professor McCanney was ahead of the rest of the scientific community by a very long shot, I would dare to say that even to day in 2012 Professor McCanney is decades ahead of the rest of the scientists. It was during the time when Professor McCanney was publishing his research that he encounter heavy resistants from the scientific orthodox community which does not surprise me a at all.

Because of his work and research Professor McCanney was sacked from his teaching work which if you think seriously now the behaviour of those individuals that sack him was a form of Neo Nazism which its really against Democracy but more importantly is not reason the foundation of Democracy? And of the people responsible for Professor McCanney sacking are ruled by greed then is it not logical to expect that greed will destroy reason? In fact what followed latter after his dismissal he was prevented from publishing any more of his work. But does it not surprise you that even if some people don't agree with his theories, are not scientist suppose to keep an open mind? And even work together to see if his work are correct rather than behaving like a lynch mob in one of those western movies? Think for a moment all those inventors in the past that had made a difference in the world we live today they came up with ideas that are extraordinary and yet it seems that if you came forward as Professor McCanney has done in research that could revolutionise our space program instead of helping him they instead want to lynch him, this is not good enough folks.

Thank God Professor McCanney never gave up, because for the next thirty years he has continue on his work and research and he never gave up his power to others, and what he has done has been to built on his work by collecting further data from the space agencies and from the planetary encounters that the spacecrafts have had and has managed to work how cosmos work's including the formulation of how the Earth weather works, a huge undertaking the he has done is just enormous. Professor McCanney has also worked out contrary to the NASA Alice in wonderland Fairy Tales that comets are not dirty snowball, and he fact findings proves that comets are electro magnetic phenomena in the cosmos and at the present time he is working on the Electro Magnetic Propulsion System this sound like the best form of space travel and its something that I hope should receive serious attention if we are indeed serious about the Space Programs, I also hope that politicians would do well in listening to Professor McCanney and to support him by funding his projects and research. Because think on the alternative, antimatter, rocket fuel and nuclear rockets are not the answer here because sound reason should tell you that those sort of spacecrafts will never be able to achieve the speed needed to fly fast enough into other planets in our cosmos and he is working on now on how do we move about in the cosmos using the electrical and magnetic propulsion system? Which is a very interesting question when you think about it.

Professor James McCanney became aware of the electrical currents and the electrical fields of energy he describes them as huge rivers throughout the cosmos, those rivers are connected from planets to their sun and to other suns, which goes all around the universe. Professor James McCanney believes that the day will come when technologies will reach the point when we will learn how to utilise those electric rivers/oceans of fields in the cosmos and we will be able to navigate on the seas by using the energies of the existing wind of the cosmos. Based on Professor James McCanney theories it would be very possible for our future astronauts to travel on spaceships using the rivers of those electrical currents and the electrical fields of energy, imagine that and you would not even need any fuel rockets or antimatter engines instead you could travel our cosmos on the energies that are already there.

Massive Comet Impact On 2011-10-01

There was a massive Impact that took place on 2011-10-01 last year. This comet was discovered by amateur astronomers however what you may not know was that from the 1 and 2 of October NASA pulled the plug on SOHO C2 and C3. When I watched this on C2 and C3 to me it look like this comet went out with a massive explosion and disintegrated when it plunged into our sun then suddenly there was a sudden explosion on the sun surface creating a massive corona ejection just after the comet impacted on the sun. Are the comet impact and the huge corona CME link together? Or was this just another bizarre coincidence? According to the orthodox scientific view the comet in question, was just a very bright comet but was not a large comet at all, and the massive CME following the comet was just a coincidence. However According to Professor James McCanney he has a different view regarding this events of October 1 2011 in which NASA pulled down the SOHO satellite site and they closed it for two days. So far NASA has not provided me any answer regarding this incident and I am still waiting for their respond. According to Professor McCanney

The activities on the sun was low and on Friday he declared that there would be a comet coming in towards the sun from the South and he stated that this comet would hit the sun in twelve hours. At that instant NASA pulled the plug on SOHO C2 and C3 and NASA brought it back online again until Sunday which means that the SOHO satellites was shut down for two days. Finally Professor McCanney managed to obtain two some photos of the comet impact and in his owned words "the sun went absolutely ballistic" when this comet hit the sun.

Interesting to note that the, space weather dot com posted some informations about our sun having two CME`s but they never wrote anything regarding this comet. Only after huge pressure after the weekend did space weather dot com said that it was just a coincidence. But this story gets a little more strange the more you research NASA sites, we are still on the same subject mind you, however the dates changed and for some reason NASA has not bothered to update their news despite the fact that I did wrote to them informing them of a mistake that they had done concerning the date.

Here is the site the title is, Incoming Comet; Outgoing CME *http://www.nasa.gov/mission_pages/sunearth/news/comet-cme.html*

This was written by Karen C. Fox from the NASA Goddard Space Station. She writes as follows Quote: On October 2, 2011, an exceptionally bright comet headed toward the sun and disintegrated. Moments later a large coronal mass ejection (CME) blew off the other side of the sun, making for this captivating movie from the SOlar Heliospheric Observatory (SOHO).

While it looks to the casual observer that the comet triggered the ejection, the apparent relationship between an incoming comet and a CME is only a coincidence. At this stage of the solar cycle, the sun is producing many mass ejections—in fact there were several earlier in the day—and it is only chance that one of them burst off the sun at the same time the comet approached. Some researchers have been looking for a more direct relationship, but nothing has yet come out of these efforts.

The comet shown here was a comet known as a Kreutz sungrazer. When a comet comes this close to the sun, it is almost always destroyed—we see the comet going in, but not going back out. Unquote. Notice the date of the comet impact is wrong it was not on the October 2, 2011, it was on October 1, 2011

Comets Reacting To The Sun

The theory of the comets by the orthodox scientists has been that it is a dirty snowball and when it gets close to our sun indeed to any sun, the sun then heats up the comet and those water and dust is then emitted from the nucleus of the comet and the solar wind the starts to blow it away producing those very long tails that you would see when you are in the field watching the night sky or on films or photographs. What that theory does not explain is how is it that if a comet is just a small snowball the comet should had vaporise long time ago, shouldn't it?

According to Professor James McCanney he has a very different theory which has been proven to be correct, is that a comet would head to this electronic field which surround the sun. Professor McCanney claims that the sun is surrounded by a negative charge and the Ion cloud is a positive charge so as a comet nucleus comes in towards the sun the sun hit's the nucleus of the comet with a an electro beam which is invisible as its started to integrate with the dust and gasses it will start to emit X rays. In simple terms he explains it in this way, it's the same principle to a Killer Fly Zapper as the insect gets closer to the light there is a space that is between the wire mesh device, so what happen is that the insect receives a high voltage electric current and it kills it. This is the same principle that Professor McCanney uses on the comets.

Also regarding our planets in the solar system according to Professor McCanney planets are also discharging capacitors which changes our weather all of this are related he says to the electronic nature of our solar system. The fact that a lot of X-rays activities has been detected from comets and just in case you are not aware X rays are a side affect of high voltage.

For more information about this subject and others I like to suggest the following books by Professor James McCanney

Planet-X, Comets and Earth Changes
Principia Meteorlogia: The Physics of Sun Earth Weather
Altantis to Tesla
The Diamond Principle
Surviving Planet X Passage
Calculate Primes Direct Propagation of the Prime Numbers The Oldest Problem in Mathematics has Finally Been Solved

OUR SUN

So you ask what is the sun made of and how hot is it? You will note that I mention that the sun has a surface in fact according to our science to date our sun is just a giant ball of gas at least this is the present theory but perhaps in the future this theory might change again in time just like evolution theorise that are accepted today as fact could be prove in the future to be wrong. If you know your history did you encounter that between four to five hundred years ago scientists such as those of today scientists and astronomers believe that the Earth was flat?

So imagine our sun to be an orange and you slice this orange in half imagine for a moment that you are observing the half of the sun that you just cut and in the centre is what its called the Core, next is the Radioactive zone, Connective zone, Granulation cells, Photosphere, Chromospheres, Corona, Flares and prominences and Sunspots which are on the photosphere. When looking at the sun you will see that the sun looks very different depending on the different wavelength that you may be using, **however a word of caution if you wish to observe the sun please wear proper eye protection and if you are viewing the sun through a telescope always use the appropriate lens filter and always check that the filter is not damage.**

The visible part of our sun is the one that you see every day and its particular beautiful during sunset as well, this part of the sun is by far the coolest but still the temperature is a massive 6,000 degrees which is cause by the sun emitting the radiation of our sun also you will note that this coolest part of the sun is the only part that can be seen with our eyes, Then as you view the sun on a shorter wavelength the temperature of the sun is more massively higher, for example the Chromosphere is seen viewed in

Altra-Volet wavelength and the sun looks red and the temperature is a huge 20,000 degrees, and when viewing the sun on the blue colour spectrum the temperature here is enormous at 800,000 degrees and then when you get to viewing the sun on X-Rays wavelength which is the Corona the temperature here is a colossal 2,000,000 million degrees. What this means is as I mentioned before is that the sun looks very different depending on the different wavelength that you are viewing our sun, but what is of critical importance here is the temperature of the sun that we are able to detect using this sort of scientific research. Now the Corona is what I am very interested on because it's is massively hotter than the surface of the sun and the Corona is even more massively hotter than the Photosphere. One other point that I want to make is that the sun has very huge bright areas and they are called, Solar Active Regions in those regions the magnetic field are extremely powerful and it is precisely on the Solar Active Regions that the Corona temperature are colossal it is also in this region that we see those massive CME, and gigantic Solar Flares explosions. It was in the Corona region that comet lovejoy flew by at 120,000 kilometres and those scientists expects us the public to believe that a little microscopic bit of ice and snow can survived such colossal temperature? It does not take a scientist with Doctorate degrees or PhD to know that such a story is nothing less than comical and yet this is the fairytales that our children are learning at school, that comets are just dirty little snowball.

Spacecraft Giotto

Between march 13 and 14 in the year 1986, the European spacecraft Giotto took a few thousand pictures to be exact it took some 2,333 photos of halleys comet. Those photographs were taken before the closest approach of around 596 km on march 14 of 1986 what those photographs shows were astonishing, for so many years hundreds perhaps from the beginning of the formation of planet Earth this comet has been coming and going has always been a site of immense beauty so were are the ice, water, and snow that halleys comet is suppose to have? The photos of Giotto spacecraft reveal something totally different to what we have been taught by our schools and teachers, why you think that is? The surface of Halleys comet resembles a rocky and dry dark surface.

So why is NASA telling the public that the comets are just dirty snowball? Do you need to have a PhD to know what a comet looks like? Lets continue and see.

Comet Tempel 1

Image credit NASA

In this picture you will see the same comet which is called Tempel 1. Can you tell which one of this photos shows the real Tempel 1 comet? Is it the one on the right hand side? Or is it the one on the left hand side? Better still can you explain what is wrong with this photographs? The comet on the left side is the real one but the one on the right hand side is a picture of what NASA was expecting to see and they failed and failed miserable because as I said before you don't need a PhD or doctorate degree to observe for yourself that the real Tempel 1 comet has nearly no water and the expect ice and water gushing fissures way out into space to enable a comet to produce those enormous coma which is the head and those massive long tail is no were to be found. Tempel 1 comet is just a dry dark object and the dirty snowball model just does not fit. But raise your hand in a classroom and ask your teacher what a comet is made off and he/she would tell you some sort of fairytale that goes something like this, comets are loosely-bound aggregates of ice and dust that form a coma and a tail of dust and plasma. Which of course is not correct for have your teacher seen Tempel 1 comet above they would see that comets are not made of ice.

The idea that a comet was just a little dirty snowball was brought about by Dr. Fred Lawrence Whipple many years ago more on Dr. Fred Lawrence Whipple latter. He suggested that a comet was made mostly of ice water and that the surface of the comet was covered by 100% this beleve was held up to 1985. In fact the late Carl Sagan co-with his wife Ann Druyan which was entitle "Comet" and on plate X, it shows a picture of a comet which is 10 kilometers across and shows just before it blast the Earth which according to their theory kill off all the dinosaurs and created a global winter. The painting in question was made by Don Davis which shows bright snowball ten kilometres in diameter, so this theory that comets were 100% ice and water was widely accepted even by the experts of those days long gone.

Between 1986 and 1987 reduced this theory to 60%, by the year 2000 and 2003 the expected ice surface theory fell to 40%. But as you can tell by the above photograph there is hardly any water or ice on Tempel 1 comet so how is it that today's scientists are still holding on to Dr. Fred Lawrence Whipple theory that comets are little dirty snowball?

The findings of DEEP IMPACT spectrometers is prove that comets are more than just Whipple theory but instead scientist are continuing with Whipple laughable dirty snowball model which has been proven not to be so by the evidence of the

DEEP IMPACT spectrometers on Tempel 1 comet shows the following: approximately 12 billion square feet of surface area of this comet nucleus, only 300,000 square feet indicated any presence of water and this was sparse and mixed with dust. However when you look at the photo above you will not see any water coming out of those areas that would be needed in order to create the tail also the small amount of water should had gone long time ago. This information was posted by the journal SCIENCE however those publications are not available to the general public, you need to subscribe to them in order to read those statements.

I have included the website in case you are interested *http://www.sciencemag.org/*

Comets has always been a source of wonder and majesty through out our history it has been believe to be both the bringer of good omens as well as the bringer of bad omens. Even during a solar eclipse when all the stars and planets becoming visible again comets have been known to be seen during this totality. This occurred in more recent times such as Comet Tewfik which came into view during the 17 May 1882 eclipse, Comet

Rondanina Bester was also seen on 20 May 1947, Comet C/1948 was also seen during the eclipse of 1 November 1948, Comet Hale Bopp also came into view on 9 March 1997. However with today's Stereo space craft Scientist have been able to detect even more comets during total eclipse such as comet Bradfield on 16 April 2004. Stereo are just two spacecrafts that were launched at the same time these two spacecrafts are not orbiting the Earth instead they are orbiting our sun in an area called the, heliocentric drift orbit. Its has been several years now since these two stereo spacecrafts were launched from Earth and are now orbiting around our sun, but those two stereo spacecraft unlike the satellites that orbits our planet earth, those two crafts are orbiting the sun with the earth. To put it simply One is orbiting on the right hand side of the sun and the other is orbiting on the left hand of the sun. The name stereo is shortened for Solar Terrestrial Relations Observatory.

The spacecrafts were launched on Wednesday 25 of October, in the year 2006, from the Cape Canaveral Air Force Station in Florida U.S.A.

The time of the launch was 8:52 PM Eastern Daylight Time, the Stereo mission was only suppose to be for only two years which was completed two years ago, but Stereo mission has been stretched and in all that time the Stereo Mission has provided lots of data and information's on the CME`s shortened for Corona Mass Ejections. This CMEs are causing great climate changes on our planet Earth as well as in every planet and moons in our Solar System including our former planet Pluto and its Moons.

How is it that our outer planets are experiencing climate change? And more importantly what its causing our sun to emit such massive explosions?

These Corona Mass Ejections phenomenon are all watched by Instruments that are named Corona Graphs which are telescopes that creates those artificial eclipses which you see on SOHO shortened for Solar Heliospheric Observatory sites they are the same as when you are watching a solar eclipse its just the same with the Corona Graphs Telescope.

Since our Moon is 384,403 Kilometres—238,857 Miles from the Earth, it's the only natural satellite and its has been a great companion to our planet.

In ancient times the Moon was associated with the woman's menstrual cycle as well as agriculture and the effect that our Moon has had on the seeds, and the growth of plants. Also Womanhood and Motherhood became associated with the, Moon Goddess since the Moon had three main aspects, the waxing, Full Moon, and the waning and the woman's life also can be shown to be seen as having this three major greatness such as pre-fertility, fertility and menopausal.

So the Moon Goddess was very much perceived as having three major stages, the Virgin, Mother and Crone. This three stages are also all represented in our human life cycles, Birth Marriage and Death, also like womanhood the Earth also nourishes of the seeds which is also associated with the Goddess.

Since the Moons diameter is 3,476 Kilometres wide, 2,159.88626 Miles wide the diameter of our Moon is a perfect match when we are viewing a solar eclipse.

Edmund Halley

Of all the comets known the most famous of these and is one of the smallest is comet Halley's, or better known as Halley's Comet this comet derives its name by the fact that it was discovered by Edmund Halley in 1682. it's the first known comet to be acknowledge as being a periodic. It was Edmund Halley that predicted in 1705 that the comet would be seen again over London in 1682 and would appear again in 1759 which meant that this comet had a periodic orbit of 75 years or so.

Edmund Halley also stated that this same comet was the one that was seen in 1607 and 1531.

The comet did in fact appeared in 1759 just as Edmund Halley predicted, who unfortunately He died in A.D. 1742 and did not live long enough to see his calculations verified. He was known as a Mathematician, Astronomer as well as an Inventor.

When Halley's comet appeared in 1910 there was a great fear put forward it had nothing to do with superstitions as the mainstream scientists have been stating in all their books and lecture. In fact the only reason why this comet was not much of a danger its because it is far too small to be of any threat to Earth but in the 1880`s scientists had a very clear understanding of what comets were.

When Halley's comet appeared in 1910 there was a great fear put forward but the fear in man's heart at the time had nothing to do with superstitions as the mainstream scientists of this century have been stating in all their books, lecture and so on. In fact the only reason why this comet was not much of a danger its because it is far too small for the poisons' to be of any threat to the Earth but in the 1880`s scientists had a very clear understanding of what comets were made of and that was why they regarded the passing of comets as a very real threat.

On page 105 The Consequences To The Earth This spectrum refers to two comets it reads as follows and I quote: ets burned with substantially the same spectrum as that emitted by burning carbon. The inference is irresistible that these comets were wrapped in great masses of carbon in a state of combustion. This is the conclusion reached by Dr, Schellen.

Padre Secchi, the great Roman astronomer, examined Dr. Winnecke's comet of the 21st of June, 1868, and concluded that the light from the self-luminous part was produced by carburetted hydrogen.

We shall see that the legends of the different races speak of the poison that accompanied the comet, and by which great multitudes were slain; the very waters that End of Quote.

The Day The Earth Stood Still

From the earliest time in recoded history people have always been frightened by this mysterious neighbours passing to our solar system. In fact according to legends from across many countries across the world including Bible stories, they all speak of great catastrophes which are extremely large scale disasters which have been accompanying after observing those celestial objects in our skies.

I am not referring to the era of King Edward VII of the United Kingdom and Emperor of India (Titles given to British Crown Rulers in those time) He died in A. D. 1910, to the political madness of the time resulting in the sudden war of World War I which started in 1914, the superstitions believes surrounding the Halley's Comet was such that all the bad omens were associated with it.

In reading Exodus 14:21-31 It reads as follows I quote:

Moses held out his hand over the sea, and the LORD drove the sea back with a strong east wind. It blew all night and turned the sea into dry land. The water was divided, and the Israelites went through the sea on dry ground, with walls of water on both sides.

The Egyptians pursed them and went after them into the sea with all their horses, chariots, and drivers. Just before dawn the LORD looked down from the pillar of fire and cloud at the Egyptian army and threw them into a panic. He made the wheels of their chariots get stuck, so that they moved with great difficulty.

The Egyptians said, The LORD if fighting for the Israelites against us. Let's get out of here The LORD said to Moses, "Hold out your hand over the sea, and the water will come back over the Egyptians and their chariots and drivers.

So Moses held out his hand over the sea, and at daybreak the water returned to its normal level. The Egyptians tried to escape from the water, but the LORD threw them into the sea. The water retuned and covered the chariots, the drivers, and all the Egyptian army that had followed the Israelites into the sea; not one of them was left.

But the Israelites walked through the sea on dry ground, with walls of water on both sides.

On that day the LORD saved the people of Israel from the Egyptians, and the Israelites saw them lying dead on the seashore.

When the Israelites saw the great power with which the LORD had defeated the Egyptians, they stood in awe of the LORD; and they had faith in the LORD and in his servant Moses. End of quote

What actually happened to the people of Israel in the red sea? Was this a divine intervention to save the people of Israel? or was the parting of the red sea a massive tidal wave cause by some celestial object? What about the trembling of the Earth that took place from Mount Sinai? Many years after how could Joshua force the walls of Jericho to collapse allowing the Jews to take the fortress so easily, by divine intervention? By blowing trumpets? Or by huge quakes created by some celestial object orbiting near our planet? When Jericho site was excavated it was found that Jericho was destroyed by an Earthquake.

In reading Joshua 10:13 It reads as follows I quote: And the sun stood still, and the moon stayed, until the people had avenged themselves upon their enemies. Is not this written in the book of Jasher? So the sun stood still in the midst of heaven, and hasted not to go down about the whole day. End of quote

What cause the sun to stay still? Is it possible that the cause of our Earth was because the rotation of our planet was slowed down? But if this was the case, what was the cause of our Earth slowing down? But if this was the case there has got to be other evidence to prove that the Earth did indeed slowed down, would it?

It is reported in the Chinese chronologies that in Yahous time, I quote: the sun did not set for a number of days, the forest were set on fire, a high wave, reaching the sky, poured over the land. End of quote

In Tractate Sanhedrin of the Talmud it is said and I quote:
Seven days before the deluge, the Holy One changed the primeval order and the sun rose in the west and set in the east. End of quote.

Even the ancient Egyptians had stated in their writings that the suns above changed four times, and that the sun has been setting twice in the part of the skies were the sun rises today.

So as you can see from those catastrophic events has had a deep impact in our entire planet and this is why I have been making a point that that those scientists are completely loosing the play here and are just making fairytales stories on both the Oort cloud myth and the Idea that comets are just little dirty snowball just does not add up.

Massive Comet Impact On October, 1, 2011

The particular comet which Impacted our the sun on October, 1, 2011, hit the sun followed by a massive explosion which must has shook the sun of its socks followed by a huge CME the size of which indicates that the massive explosions cause by that comet could not had been cause by a little dirty snowball as Nasa and other orthodox Astronomers keep repeating,. Is it a fact that following my own observation on this comet is it not a weird thing for Astronomers continue to follow this error that comets are just a few miles or kilometres across and are just a little fuzz ball if comets are not snowballs are there any prove to determined what they really are? also was the massive explosions of the CME that followed after the comet hit just another coincidence? As astronomers keep suggesting?

Or could there be a linked between the massive comet hitting the sun that was immediately then followed by those huge explosion resulting in massive amount of CME`s?

And why did Nasa had the Soho closed down for two days? Is there a link between the comet hitting the sun and Nasa closing down Soho? If so Why?

NASA Shuts Down The Space Shuttle Program

Its a real pity after all the money that the American tax payers are putting into that organization its just a shame. One wonders were all the taxpayers funded money are going, and now the NASA space shuttles are no longer, all you will be left is just a public relations department doing their best to disinfo the public, but why? What do they have to gain by closing down the NASA space shuttles programme? Even Texas Gov Rick Perry was very upset at the Obama administration for closing down the shuttle flights program. According to the guardian news paper it claims that 2,000 workers would loose their jobs once the Atlantis returned back to Earth which would had meant then that the entire amount of people loosing their jobs would total around 10,000 workers give or take. Now

I am no politician and thank the gods for that but those are people with families that they need to take care of homes with mortgages that has to be paid. So what's the future going to be for the space program?

Ancient Sumerians Contributions to Planetary Science

We must not forget also the contributions that the Ancient Sumerians played in our understanding of the Cosmos. The Sumerian also kept records of our Solar System depicted in their Akkadian cylinder seals which the Late Dr. Carl Sagan confirmed that those circles are Celestial spheres which include the representation of our Sun and Planets. including also The Sumerian also kept records of Halley's Comet inscribed in their tablets as well as the planets around our solar system.

In fact Dr. Carl Sagan confirmed that those circles are Celestial spheres which include the representation of our Sun and Planets. But I know what you are going to ask me yes? How can this be you ask since the idea of planets rotating around the sun is suppose to be the idea of Copernicus. That would have been because the Greece copy learned of this idea from the Sumerians of 6,000 years before.

One of the Oldest Sumerian cuneiform tablet designated BM 41018 dated in 87 B.C. In the 1980s Hermann Hunger identified a fragment of the ancient Sumerian text that referred to a comet, later in the year 1985, F. R. Stephenson, K. K. C. Yau, and Hunger were able to use their available astronomical data that they had at hand on the same Sumerian tablet to establish the year as—86. They were also able to establish that

Halley's Comet Return July 28, 2061

Halley's Comet was seen "day beyond day" during the lunar month of July 14 to August 11, and that another observation on August 24 reveals the comet had a tail 10° long.

Halley's Comet have been our faithful companion since the creation of our solar system came into existence and its periodic orbits has been so precise throughout the many centuries that both past astronomers such as Edmund Halley and today's astronomers have been able to calculate when Halley's Comet would be seen in our neighbourhood once again that the next arrival this friendly messenger of our cosmos will be on July 28, 2061.

Our children and grandchildren will be able to observe this beautiful beacon in the darkness of space.

Assuming offcourse that we as a species have not destroyed ourselves with senseless wars, and instead to help humanity solve our hunger and famine instead. Given that this small comet has been around for hundreds or thousands of years you would think that if indeed as David Morrison who is the senior scientist at NASA has been telling everyone, that comets are just little dirty ice snowball, you would think that this comet would had indeed all the comets entering close to our sun should had melted by now

Instead they just keep coming in to our solar system glowing does massive long tails and coma and yet there is not water, no ice, and they are definitely look more like meteor and in those photographs they look to have a hard surface and its does not look anything like snowball.

All in all Halley's Comet has been documented thirty times as far as I am aware, however it may be that in the past this comet would had been seen by those civilisations which are now long gone. The last time that Halley's Comet comet appeared was February 9, 1986, prior to 1986 it was also observed on April 20, 1910, November 16, 1835, March 13, 1759, September 15, 1682, October 27, 1607, August 25, 1531, June 9, 1451, November 9, 1378, October 23, 1301, October 1, 1222, April 22, 1145, March 23, 1066, September 9, 989, July 9, 912, February 27, 837, May 22, 760, September 28, 684, March 13, 607, September 25, 530, June 24, 451, February 16, 374, April 20, 295, May 17, 218, March 20, 141, January 26, 66, October 5, 11 B.C. August 2, 86, October 5, 163, March 30, 239 B.C.

Short Period Comets

For such a little comet Halley's Comet it's the most famous of the short period comets also known as Helley Family. but there are others, a short period comet are those that as the name implies, it means a comet that returns back once every 200 years or less.

One thing you need to know and understand is that the list of groups both long and short periods are growing due to the two stereo spacecraft that we have orbiting around the our Sun. Some of the known comets that are short orbit period are, Oterma, 27P Crommelin, 21P Giachcobini-Zinner, Whipple, 2P Encke, 19P Borelly, 9P Tempel 1 comet is of special interest since data was taken by DEEP IMPACT and the results was that there is hardly any moister in this comet, all of this comet surface is just a hard solid surface that resemble a rock and not the, derty snowball myths that we have been told by the NASA officials. The other matter is that NASA has produce very few photograths about Tempel 1 but we hope that they will release more photos from not just this comet but of all the comets and preferable showing the surface as well.

Tempel 2, 6P d'Arrest, Perrine-Markos, 55P Tempel-Turtle, Temple-Swift, 73P Schwassman-Wachmann 1, Kopff, Turtle-Giacobini-Kresak, Honda-Mrkos-Kresak, 81P Wild 2, Arend-Rigaux, Faye.

Missing Lunar Orbiter Tapes Found

I like to thank Mr. Mr Henry A. Eckstein for giving me permission to show this evidence to the public of the mission lunar tapes in which NASA senior scientist Dr, David Morrison has written to me saying that he knows nothing about it, which is very strange given that his office is just a few minutes away from the former McDonald restaurant. It sounds comical and you have probable never even heard this story but I can assure you its all true so buckle up as you read his full story.

Lunar Surface

Here is a snip of film created from the lunar orbiter data files. Some of the files were used to create high resolution maps of the moon for Apollo. Only a very limited [number].The high resolution images have never been seen by the public because at the time, they were classified revealing the extreme precision of our spy satellites). Instead, all we have ever seen are the grainy photo-of-a-photo images that were released to the public. Hummm . . . exactly what I you are seeing here as I took a photo of the film that our host was holding up to the light. I tried hard to capture some of the lunar images on the fly =) Lunar Surface(strip)

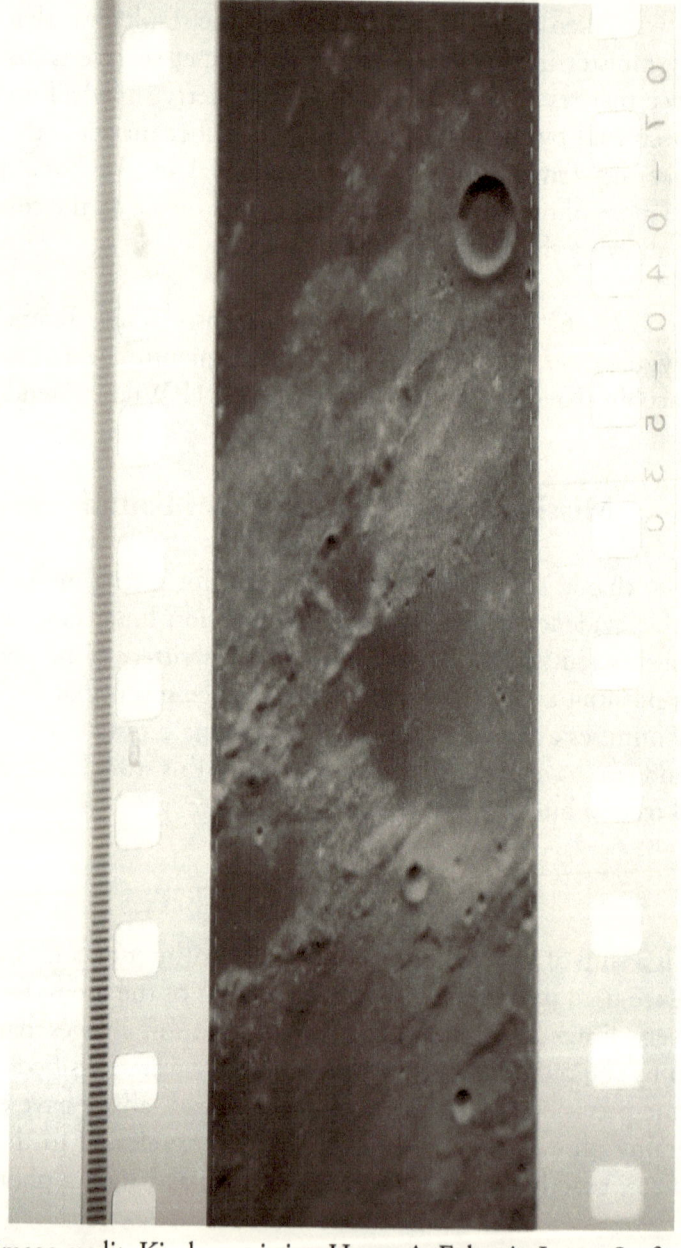

Image credit: Kind permission Henry A. Eckstein Lunar Surface

Letters by Henry A. Eckstein

Dear Mr. Correa,

I am sorry that I have taken so long to respond to your email.

I have been away for some time due to work-related perogatives and was unable to respond to your email in a timely manner.

I am speaking to you PERSONALLY as a qualified technologist in the area of video production techniques and computer graphics design and production and NOT as a representative of my workplace nor as a representative of NASA or any other government entity.

I became involved within the NASA MOON TAPES discussion on ATS (www.AboveTopSecret.com) about October 2008 wanting to help the ATS community after hearing about the possibility of recovering video from the 1969 Lunar Orbiter Video originally recorded on 2" Ampex video tape machines.

From what I understood at the time, the image data was from a moon-to-earth broadcast transmission picked up by an Australian Relay Station and thus recorded and preserved in a format called PCM (Pulse Coded Modulation) which is technically a DIGITAL VIDEO FORMAT. The original tapes were eventually shipped to the United States and stored in various office buildings within the NASA beauracracy until finally ending up in an old abaondoned McDonalds in Florida which was eventually discovered by a retired aerospace personnel who wanted to see what was on the tapes.

Given permission by NASA to do the conversion work on their own, they TRIED to find, repair and use old video gear which worked very well up to a certain point.

From what I also understand the Moon Video originally seen worldwide was VERY DIFFERENT from what was seen by the Australian Satellite Relay Station.

That station recoeved crystal clear video directly from the cameras which I understand were in a format that is the equivalent to about 600 pixels by 400 pixels Greyscale image frames.

This greyscale imagery was stored on a line-by-line basis onto 2" Quad-track video tape machines of 1960's vintage by a technique of converting the greyscale pixel data to ON/OFF pulses which show up as blocks of black & white lines and squares on video displays.

As time moved on, the old machines to read this data became harder and harder to obtain and use for the NASA recovery efforts and thus after reading about the video conversion effort on an ATS discussion forum I decided to offer my help by outlining a method which can recover digital or analog audio and video data from vintage tape reels or other media using high resolution optical cameras which convert video frame data into the digital ones and zeros that can be shown as colour or black and white imagery WITHOUT having to have access to ANY type of old audio/video tape deck.

For a technical explanation as to the NEW type of conversion system, these are the steps:

1) Obtain old audio or video tapes.
2) Place tapes into spindles containing a playback/feed reel and takeup reel.
3) Place a moving 2048 pixel or 4096 pixel camera with microscope-like lens to focus on and scan the entire width of the tape as it winds from the playback/feed reel to the takeup reel. Since the tape is usually 2 inches wide, the hi-res camera must move up and down to capture the entire width of the recording side of a video or audio tape on a block-by-block basis.

4) Convert each scan to a 2k-by-2k or a 4k-by-4k (k = 1024 pixels) 8-bit or 16-bit greyscale image and stitch multiple scans together using software to create a photo swatch that is about 4k wide by 16k pixels long.
5) Send each stiched-together photo swatch to a software application called a Digital Signal Processor which will analyze this photograph of many groups of magnetic particle clusters which are the metal oxides that form the videotape or audiotape recording surface. These clusters of magnetic particles are edge-detected using a technique called Horizontal and Vertical Sobel Convolution Filters which will determine the polarization (i.e. orientation) of the grouped iron or other metal oxide particle clusters which form the video or audio data.

We analyze all these oxide clusters together and organize them into individual lines and frames of video pixels or audio samples which represent PCM (pulse Coded Modulation) codes which will get further interpreted by my software into digital ones and zeros which can be converted to modern-day digital 30 frames per second RGB video and/or 16-bit 48kz WAVE-fromat audio data.

6) The beauty of this method is that because there is NO contact with the tape surface, it cannot be damaged AND I can filter out audio & video noise and interference using techniques from the world of mathematically-based digital signal processing. This means the NASA Moon Videos can be seen in their ORIGINAL glory and not that grainy, blotchy and nearly unwatchable video that was seen by almost everyone else in 1969.

7) Convert PCM data to 720-by-480 pixel Standard Definition DVD or 1920-by-1080 pixel High Definition Blu-ray audio/video data and post to Youtube!

The above was my original plan to offer my help in video conversion . . . HOWEVER . . . I was a victim of my own self-serving attitude which basically TRAMPLED on other peoples desires, wants and needs. I was blinded by my own ambitions and accidentally stomped on a few toes and was thus rebuffed by those who would rather do the video conversion work themselves and thus claim their rightful glory.

See these links to the original Lunar Orbiter Tapes discussion thread on ATS:

http://www.abovetopsecret.com/forum/thread394867/pg1

http://www.abovetopsecret.com/forum/thread394867/pg15

http://www.abovetopsecret.com/forum/thread394867/pg17

Look for any responses to and from StargateSG7 (aka Henry Eckstein)

So in terms of my involvedment, there was NONE on my part other than adding a colourful addition to the ATS lunar orbiter tapes discussion thread.

From what I understand, the conversion process of teh NASA Lunar Orbiter tapes is still not yet fully done and I STILL BELIEVE (as of January 2012) that my conversion method WOULD BE of superior video and audio quality AND it would have been done SO MUCH faster. With the estimated 1500 tapes I thought were still out there, I could have done the conversion in a few weeks using moving optical scanners ten-tapes-at-a-time onto a few hundred DVD's or tens of multi-gigabyte hard disks.

Letters by Henry A. Eckstein

Dear Mr. Correa,

While I personally have had nothing ELSE to do with the processing or transfer of audio and video material residing on the Lunar Orbiter Tapes found within an abandoned McDonalds in Florida, I do have some general knowledge about U.S. Military programmes dealing with proposed and possibly now active High Orbital and/or Lunar Bases (i.e. both manned and unmanned) used for both scientific evaluation programmes and for research and development into systems used for the Strategic Defense Initiative programmes (SDI) that were initiated by Ronald Reagan starting in 1983 and continuing even now.

In the following paragraphs, I do rely heavily on Wikipedia entries to give you an outline on particular systems, but there are ways to use Google Search to obtain specific reports made public by DOD agencies and others who have an opinion on and have specific facts about POSSIBLE Lunar Bases and Orbital Weapons Systems. It would be your perogative to do some extensive searching to gather these open-source and private documents.

Some Historical Perspective:

During the 1950's and early 1960's, the then administrations of Harry Truman (April 12, 1945 to January 20, 1953) and Dwight D. Eisenhower (January 20, 1953 to January 20, 1961) were concerned by Soviet (USSR) advances in rocketry and Nuclear Weapon Payloads that could be used for space-based deployment that could threaten the U.S. Mainland and the territories of its allies.

While publicly decrying the use of space-based weaponry, both the USA and the Soviet Union secretly began the evaluation of technologies and systems that would allow just such a deployment of kinetic energy-based and nuclear payloads into Earth orbit AND onto extraterrestrial environs such as the Moon and even Mars!

Since the 1990's, numerous details have arisen about these programs that, on a general basis, were abject performance failures in actually getting anything into orbit or onto the moon. Declassified reports and participant anecdotes have recounted the trials and tribulations of testing rocket-based vehicles and orbital deployment systems that either could not be built because of financial considerations, or were abandoned due to then technological problems with specific deployment and upkeep technologies that would be based within orbital and off-Earth environments.

The following weblinks should give you some idea of what was possible in the 1950's and 1960's.

See the following Weblinks:

Project Horizon, A U.S. Army Study for the Establishment of a Lunar Military Outpost.

Project Horizon Moonbase (1958):
　http://en.wikipedia.org/wiki/Project_Horizon

Military manned lunar landing prior to the Apollo Program (USAF 1958):
　http://en.wikipedia.org/wiki/Lunex_Project

Project A119: A Study of Lunar Research Flights:
　http://en.wikipedia.org/wiki/Project_A119

Soviet Union Zvezda moonbase DLB Lunar Base Project:
　http://en.wikipedia.org/wiki/Zvezda_%28moonbase%29

On a more recent note, after President Roland Reagan initiated the Strategic Defense Initiative (SDI or Star Wars) programme in 1983, sudden improvements to both computer-aided design and machining systems plus the enhancements to general computing systems allowed an ability to actually design, build and deploy an active ballistic missile defense system INTO EARTH ORBIT and allow the basing of offensive kinetic energy or explosive payload delivery systems into high orbit or onto the moon!

See Weblink for SDI (1983):
http://en.wikipedia.org/wiki/Strategic_Defense_Initiative

Strategic Defense Initiative Organization (1984)
http://en.wikipedia.org/wiki/Strategic_Defense_Initiative_Organization

Ballistic Missile Defense Organization (1993)
http://en.wikipedia.org/wiki/Ballistic_Missile_Defense_Organization

Missile Defense Agency (2002)
http://en.wikipedia.org/wiki/Missile_Defense_Agency

Kinetic Enery Weapon:
http://en.wikipedia.org/wiki/Kinetic_energy_weapon

Kinetic Projectiles:
http://en.wikipedia.org/wiki/Projectile#Kinetic_projectiles

Evolutionary Air and Space Global Laser Engagement:
http://en.wikipedia.org/wiki/Evolutionary_Air_and_Space_Global_Laser_Engagement

Deployed Theatre Ballistic Missile Defense:
http://www.army-technology.com/projects/vlmica/

In the following paragraphs, I must state that my words are personal statements and conjecture along with some "somewhat insider information/anecdotes" that may illustrate current space-based military systems that ARE POSSIBLY based in Earth orbit or even on the Moon.

Current Openly Admitted But Classified Operations Orbiting Vehicles:

USAF X-37 Small Payload Space Plane:
http://en.wikipedia.org/wiki/Boeing_X-37

Manned Orbiting Laboratory (Now Operational—Schoolbus Sized—4 to 6-man crew):
 http://en.wikipedia.org/wiki/Manned_Orbiting_Laboratory

The above is an interesting one as it was DENIED to be in operation but now seems to be a wink-wink-nudge-nudge open-secret space station which is currently deployed as an operational moveable optical spy satelllite KNOWN to Russian and Chinese space agencies BUT but having a secretive secondary purpose which allows USAF astronauts to be deployed ON-BOARD for extended periods of time in a non-observable module about the size of a small schoolbus. While cramped and diminutive in size, it nonetheless has its uses in the research, development and manufacture of crystaline substrates for ultra-high-speed radiation hardened microchips and for human-centric, space-based intelligence gathering operations. . . . AND . . . other more secretive military operations!

Other More Speculative Unconfirmed Systems:

Hardened Communications Relay System at Lunar Pole:
 http://theworld.com/~reinhold/lunarpolar.html

The above was a 1990's era speculative foray into creating a robotic unmanned lunar station that could be used for lunar environmental monitoring operations AND for use as a possible adversary-resistant communications relay station for military communciations.

Speculative Unmanned but Eventually Moon-based Mannable Space Shelter and Space Vehicle and/or Space Construction Systems:

Possible Project Code Names:
  a) Project Lunar Spade—High Orbital and Moon-based sub-surface sheltering systems
  b) Project Diamond Moon—High Orbital or Moon-based optical and non-optical telescopes
  c) Project Plastic Moon—Possible relation to orbital or moon-based composite structures manufacturing
  d) Project Copper Moon—Possible relation to a high orbital or Destination-Moon space plane

The above items are the only items I have here that is purely based upon oral anecdotes from industry insiders who MAY HAVE specific information on moveable platforms that are designed to manufacture moon-based or orbital shelters and space vehicles from moon-obtained material. Specific research was said to have been done to create silicon-glass-fibre composites from basic moon-rock and robotically or humanly mine specific non-Earth observed areas on the moon for raw metallic and mineral-based material for the construction of high-performance space planes and fuel storage systems which would take off from the Moon for Earth-centric operations . . . OR . . . be used to geosynchronously station said vehicles and shelter platforms AWAY from Earth-based eyes for future operations.

From what I understand, these above projects had preliminary feasability studies starting in 1988 and became active research and development projects by 1992 with SOME initial workable/tested systems completed by 2004-2006 and for POSSIBLE actual deployment by 2008 to 2014 timeframes. Again, this is majority conjecture and off-hand unprovable remarks but certain patents and systems development relating to robotic silicon-glass-fibre composite construction systems and techniques along with patents relating to space fuel (Hydrogen/Oxygen/Peroxide) production and storage systems has me wondering what the USAF is up to regarding their intentions on the Moon or moon orbit.

I also have minor off-hand knowledge about other more-secretive projects that you may be interested in regarding advanced field-effects based space propulsion systems and autonomous space vehicles but I should note much of that knowledge is also off-hand anecdotes from 3rd parties.

Please do keep me abridged of the progress on your book and please do email me if you have any further questions.

On a final note, I actually do have SOME non-copyrighted open-source imagery that was sent to me by outside interests but I need to search for those images regarding designs for high-orbit vehicles and space-based or moon-based structures. I shall notify with 7 days of whether I have those images and if so, I will send them to you for reproduction in your book.

Thank You,
Henry A. Eckstein

Some Discussion Taken Place at Above Top Secret

Bingo Rocketeer for spotting mission tapes and jerryfi_99 for guessing that imagery data would take this much space.

The Pirate flag is purely motivational, methinks, for a skunkworks improvising what was thought to be impossible.

Forty years ago, unmanned lunar orbiters circled the moon taking extremely high-res photos of the surface to plan landing spots for Apollo 11 onward . . . In this McDonalds, the only copy of that data is about to be resurrected. Erik and I dropped in for a visit after the LUNAR rocket launch at NASA Ames.

And gosh, Alieness may be right too when they look at those images carefully for three-toe footprints . . .

They have never been seen by the public because at the time, they were classified because they would reveal the extreme precision of our spy satellites. Instead, all we have ever seen are the grainy photo of a photo images that were released to the public.

The spacecraft did not ship this film back to Earth. Instead, they developed the film on the Lunar Orbiter and then raster scanned the negatives with a 5 micron spot (200 lines/millimeter resolution) and beamed the data back to Earth using yet-to-be-patented-by-others lossless analog compression. Three ground stations on Earth (one was in Madrid) recorded the transmissions on these magnetic tapes.

Recovering the data has proven to be very difficult, requiring technological archeology. The only working version of the Ampex tape player ($300K when new) was discovered in a chicken coop and restored with the help of the original designer. There is only one person on Earth who still refurbishes these tape heads, and he is retiring this year. The skills to read this data archive are on the cusp of disappearing forever.

Some of the applications of this project, beyond accessing the best images of the moon ever taken, are to look for new landing sites for the new Google Lunar X-Prize landers, and to compare the new craters on the moon from 40 years ago, a measure of micrometeorite flux and risk to future lunar operations.

And yes, the conspiracy continues, with McDonalds' long and sordid history with the Apollo program . . .

Behind the counter of an abandoned McDonalds lie 48,000 lbs of 70mm tape . . . the only copy of extremely high-resolution images of the moon.

These tapes were recorded 40 years ago as part of the Apollo program to map the lunar surface to plan landing spots for Apollo 11 onward. They have never been seen by the public because at the time, they were classified as they reveal the extreme precision of our spy satellites. Instead, all we have ever seen are the grainy photo-of-a-photo images that were released to the public.

The spacecraft did not ship this film back to Earth. Instead, they developed the film on the Lunar Orbiter and then raster scanned the negatives with a 5 micron spot (200 lines/millimeter resolution) and beamed the data back to Earth using yet-to-be-patented-by-others lossless analog compression. Three ground stations on Earth (one was in Madrid) recorded the transmissions on these magnetic tapes.

Recovering the data has proven to be very difficult, requiring technological archeology. The only working version of the Ampex tape player ($300K when new) was discovered in a chicken coop and restored with the help of the original designer. There is only one person on Earth who still refurbishes these tape heads, and he is retiring this year. The skills to read this data archive are on the cusp of disappearing forever.

Some of the applications of this project, beyond accessing the best images of the moon ever taken, are to look for new landing sites for the new Google Lunar X-Prize robo-landers, and to compare the new craters on the moon today to 40 years ago, a measure of micrometeorite flux and risk to future lunar operations. This project started in the late 1980's when the National Space Science Data Center (NSSDC) discovered a cache of the only known remaining set of Lunar Orbiter tapes in existence stored in a "salt mine." The story there is that there are abandon salt mines that store government records, as the temperature and humidity are stable. There was some documentation attached indicating what they were and that JPL should be notified as to what their ultimate fate should be. JPL took possession of them in about 1988 or so, as there was some interest in recovering the data so that the images could be digitized and made available to the general public as the pictures were then a bulky 2000, 28" x 30" prints. The problem at that point was that no one knew what technology created the tapes so the format and method was unknown. At the time a private consulting firm became aware of the project and decided to research the issue with the purpose of proposing a data recovery project. After amassing all the Lunar Orbiter literature available, it was determined that the Ampex FR900 tape recorder (the first real video tape recorder), was used to create the tapes. More importantly it was revealed that the data was in an analog format with the video in a format called "Vestigial Sideband Filtered", slow scan TV. This knowledge set about the search for any source of FR900 tape drives. The search covered NASA sites, Vandenberg's Pacific Missile Range at Kwajalein, the CIA and Egland AFB's radar test site in Florida. Ultimately a total of four tape drives were obtained and as far as is known, are the only remaining drives of their type in the world.

The next problem was to determine if the drives would read a tape without destroying it. After numerous calibrations and experiments on spare tape, it was determined that it would be safe to try one of the Lunar Orbiter tapes. This was done and the specified video spectrum was obtained which proved the capabilities of the drive and that the data on the tape was still there. However, in order to obtain the video from the data, a circuit called the VSB decompressor (or "restorer"), needed to be designed and constructed. This was done and a recognizable sync pulse with video data was retrieved.

This was all accomplished in about 1992. Since then several proposals to NASA and various private sources failed to produce the money required to recover this data. So the tape drives were stored in a "chicken coop" (actually it was a garage / barn combination), for the next 15 or so years. Last year a call was made form the person in the video (who I will only identify as "D" until I can obtain his permission to release his name—though I don't think this mission is actually a secret), called to ask about the tapes and the tape drives as he had some contacts that might be able to help. After visiting the "chicken coop" and ascertaining that the tapes were still at JPL's storage facility, he then made arrangements to transport both to a site in Northern California from the Los Angeles area, which he did. He then assembled a crew of experts in various fields and located a site to carry out his low budget "proof of concept" which turned out to be a McDonalds, which was located on a military base, that was closed due to poor attendance after a government cutback. As it turned out, each of the little tables, normally used for enjoying your "Happy Meal", were excellent workbenches for the various projects associated with bringing the drives back to working condition.

This then is pretty much where it is today. Once five full images are recovered, then the "Proof of Concept" will have been achieved and further funding may follow.

Someone mentioned that Google might be interested. Well they are, and they have visited the site.

Some information that might be of interest:
The products of the original LO project were:

Tape: 2" wide tape that contained all the mission data. There are about 2,000 of them

GRE file: Ground reconstruction 35mm film that were the reconstructed picture "framelett" data About 30 of these make up one of the 28" x 30" prints.

Prints: The prints are about 28" x 30" in dimensions There are 4 kinds of data on the FR900 Lunar Orbiter tapes:

Video: From sync to sync represents one scan line. There are 4 scan lines per meter on the surface at best resolution.

There are about 16,600 scan lines per framelett and there are about 30 framelett's per print. Telemetry: Another channel in the video data is for telemetry, which reports on the status of the satellite.

Carrier: In order to reconstruct the original video data the carrier must be available.

But because it was suppressed by the VSB processing, it was divided by 8 and stuck in the lower sideband.

Audio: There is an audio channel that the various sites recorded the tape ID on as can be heard in the video.

Film in refrigerator:

Some have said that storing film in a refrigerator would extend its life. This is true only for unexposed film.

For exposed film, it is best to have a stable temperature and humidity, cold has no effect.

Tape storage in canisters in picture:

Note that the canisters in one of the pictures with the yellow tape. These are the 2" wide original LO tapes and are in excellent condition. What is not in the picture are the GRE film, which were also with the tapes.

Pirate Flag:

The pirate flag was placed on the window was for fun as it was seen by some that this mission was going on oh, shall we say, by any means possible. More for humor I am sure.

Moffett Federal Airfield
NASA AMES Reseasrch Center

Moffett Federal Airfield (IATA: NUQ, ICAO: KNUQ), also known as Moffett Field, is a joint civil-military airport located 3 miles (5 km) north of Mountain View, in Santa Clara County, California, USA. The airport is near the south end of San Francisco Bay, north of San Jose. Formerly a United States Navy facility, the former naval air station is now owned and operated by the NASA Ames Research Center. Tenant military activities include the 129th Rescue Wing of the California Air National Guard, operating the HC-130 Hercules, MC-130 Combat Shadow and HH-60 Pave Hawk aircraft, as well as the adjacent Onizuka Air Force Station and Headquarters for the 7th Psychological Operations Group of the U.S. Army Reserve. NASA also operates several aircraft from Moffett, including the ER-2, a civilian research version of the U-2.

By far the most famous and visible sites are hangars #1, #2, and #3, which dwarf the surrounding buildings. Hangar One is one of the most remarkable hangars in the world[citation needed]. Hangars #2 and #3 are significant more for their size than their unique styling or design. Hangar One is a Naval Historical Monument and the entire airfield is a United States Registered Historic District.

In May, 2008 The National Trust for Historic Preservation listed Hangar One on their list of America's Most Endangered Places.

The NASA Ames site is home to several wind tunnels, including the Unitary Plan Wind Tunnel (a National Historic Landmark), and the National Full-Scale Aerodynamic Complex (NFAC). Despite its closure as an active military base, Moffett Field still has many active facilities and residents. Active military families still live on Moffett Community Housing, and the former base has several lodges which primarily house academics and students associated with the Ames Research Center. Moffett Field's facilities available to residents include a pool, post office, golf course, tennis courts, gas station, and several small shops and restaurants, including an on-site McDonald's which closed April 30, 2008.

Image Build

The team is using a workstation to integrate all of the images into a unified lunar map. Here he is showing an example of the resolution that was shown publicly. All we have ever seen are these grainy photo-of-a-photo images. The high resolution images have never been seen publicly, because when they were developed in the 1960s, they were classified (revealing the extreme precision of our spy satellites).

Question:

Hello,I am from China,yesterday, I watched a scientific program [Explore and search], which said that: two previous employee of NASA write a book named [DARK MISSION], which debunk many truth about MOON and MARS covered by NASA, such as GLASS COVER on the moon, artifical tower, even anti-grav device. Is that real? A Widely spreaded story that Astronaut of AOPPLO 11 witnessed UFO on the moon, is that real? If it is not real, who made this story?

Answer:

NAI Senior Scientist answer is as follows and I quote: I am glad to know that Ask an Astrobiologist is read in China, but I am sorry that you are being exposed to so much pseudoscience. The book "Dark Mission" is not by former NASA employees. Its authors, Richard C. Hoagland and M. Bara know nothing about how NASA and the scientific community really work. Please don't take their nonsense about the Moon or Mars or anti-gravity devices seriously. Any claims that Apollo 11 astronauts saw UFOs are just lies, as the astronauts themselves have said repeatedly. There is no evidence to support such claims. You ask who make these stories.

Mostly it is people who want to sell you their books or tapes. Also, some people apparently enjoy fooling others. And I suppose a few of them really believe what they say and write. I am very sorry if this sort thing is appearing on Chinese television, but there is no law against telling lies on TV or the Internet. I hope you will read other questions and answers on this Ask an Astrobiologist website to understand better what space scientists are really discovering.

NAI Senior Scientist
December 26, 2008
http://astrobiology.nasa.gov/ask-an-astrobiologist/question/?id=4707

UFO SITINGS BY Apollo 11 Astronauts

Has there been an Astronaut or more than one astronaut that has seen and reported a UFO? So if you are a scientist or student and you ask this question and some scientist tells you that UFOs are nonsense and its all a hoax, you because you have been told by an authority in Astrobiologist would accept his answer based on trust after all he is a scientist.

But have you ever taken the trouble of doing a little research on your own asking some one that has seen them? But then again if you found that your answers contradicts your senior scientist you would much better ignore those findings in particular if you are wanting to pursue a career in Astrobiologist, or any other professional career in Astronomy because you may wondered that by proving that your teacher is wrong many doors leading to a successful career may be closed of for you is that the case?

Dr. Richard Haines Aerospace Researcher NASA Retired
Dr. Richard Haines over 30 years amassed an astonishing 3,000 reports of cases in which pilots reported seen UFOs sightings.
He was a senior research scientist at the NASA AMES Research Centre.

Michael Smith was a United States Air Force Radar Controller Sergeant.
He recounts the time he witness a UFO in which other people saw it as well, a UFO from 80,000 feet was there for ten minutes, then descended and it drop of radar for the next ten minutes then in an instant it appeared again at 80,000 feet again then on the next sweep this same object was stationed 200 miles away and its stayed there at the same location for the next 10 minutes.

Sergeant Michael Smith was later told that if he ever saw another UFO to notify NORAD.

Retired Col Buzz Aldring, PH.D. Was the Astronaut that operated the Lunar Module on the Apollo 11 mission and he was the second man to step on the moon. This man, this officer is a Hero in his earlier year in the force he flew no less that 66 combat missions in Korea and shot down two MIG 15 jet fighters. He also served as a Gunner instructor, he served for 21 in the force and retired in 1972. and this is but the tip of the iceberg.

Dr. Aldring Also served as an Astronaut in 1963, he also served in the Nasa Gemini 12 together with Astronaut James Lovell back in November 11 the year was 1966.

Dr. Aldring also served in the famous moon landing of 1969 with another great Astronaut Neil Armstrong. I still remember those precious moments in time as if it were yesterday, to me at least it meant that we would one day explore all the other planets in our Solar system and go beyond, and to discover for our selves that what we think as fact in our cosmos we would one day learned that the Universe is not at all as what we think it's today.

So it was a shock for me as I am sure to all of you as well to learned that Dr. Aldring did indeed saw the UFOs while in space. But you ask, if Dr. Aldring did see those UFOs, why did Nasa not announced it to the public? Also why did any of the Astronauts' did not reported this facts?

According to Astronauts such as Dr. Aldring Nasa was well informed about the UFOs that the Astronauts saw and reported but Nasa decided to keep it a secret. But why would Nasa want to keep the UFOs sighting a secret for so many years?

And to this day why are Scientists in Nasa saying that no Astronauts saw any UFOs?

That's this whole thing sounds weird to you?

So my question to NAI Senior Scientist is, who is fooling who?

In fact there are other Astronauts which claimed to have seen UFOs and had to lived with this foolish secret which was thrust on them.

Retired Col Leroy Gordon Cooper, this retired officer has a most interesting Bio in fact he also recived a large number of Special Awards including the Distinguished Flying Cross Cluster including, NASA Exceptional Service Medal, The NASA Distinguished Service Medal, USAF Command Astronaut Wings, The Collier Trophy, The Harmon

Trophy, The Scottish Rite 33, The York Rite Knight of the Purple Cross, The DeMolay Legion of Honor. Plus a plethora of other awards special honors.

Retired Col Leroy Gordon Cooper played an important part in the 1959 and was selected as a mercury astronaut back in those early days. In 1963 He also pilated the faith 7 spacecraft which covered a 22 orbit mission. He also piloted the Gemini 5 mission that started on August 21 back in 1965 which is 46 years ago.

He also aestablis a space endurance record with the late Retired Captain Charles Conrad, there record was traveling a distance of 3,312,993 miles, 3 312 993 miles = 5 331 745.41 kilometers, and they accomplished that in 190 hours and 56 minutes.

Retired Col Leroy Gordon Cooper has the great Honors in being the first officer of making the second orbital flight in 225 hours and 15 minutes.

Retired Col Leroy Gordon Cooper was also the main back up pilot for both the, Gemini 12 and the Apollo X missions.

It was Retired Col Leroy Gordon Cooper that revealed the secrecy that are thrust on to him as well as other Astronauts. In fact he also stated that the USA radar picks up objects on a daily basis but none of that is ever reported due to the secrecy impose on them.

In the year 1965 onboard the Gimini Astronauts Retired Captain James Lovell and Retired Col Frank Borman reported UFOs in this incident they brock the 14 days record flight. In fact Retired Captain James Lovell reported to Gemini control at Cape Kennedy THE FOLLOWING:

Retired Captain James Lovell: BOGEY AT 10 O'CLOCK HIGH.

Capcom: This is Houston. Say again 7.

Retired Captain James Lovell: SAID WE HAVE A BOGEY AT 10 O'CLOCK HIGH.

Capcom: Gemini 7, is that the booster or is that an actual sighting

Retired Captain James Lovell: WE HAVE SEVERAL . . . ACTUAL SIGHTING.

Capcom: . . . Estimated distance or size?

Retired Captain James Lovell : WE ALSO HAVE THE BOOSTER IN SIGHT . . .

The late Retired Major Donald Kent Slayton he is another Hero he joined the Air Force in 1942 and started as a cadet to be trained in a B-25 Bomber as a pilot. This officer was a legend, he flew 56 missions with the 340 Bombardment Group. During World War II he flew in another 7 missions over Japan in a Douglas A-26 Invader which was part of the 319th Bombardment Group.

The late Retired Major Donald Kent was a very distinguished officer with Special Honors including the oldest officer to fly into space. He was also enshrined in the National Aviation Hall of Fame over 15 years ago way back in 1966.

Also the, Deke Slayton Memorial Space & Bicycle Museum in Wisconsin was also named after him to honor his memory.

So Retired Major Donald Kent did an interview some time ago in which he stated that he had seen a UFOs.

In fact the lists of professionals people is so long that If I added them all up it would take almost all of the book but I am just making the point that these Officers know the differences between a plane and some other alien craft and given also there expertise in their field I believe that their statements are true and factual.

The late Captain Joseth Albert Joe Walker who was awarded the Air Medal with seven Oak Leaf Clusters, including the, Distinguished Flying Cross.

If you have seen his photos he reminds me of the movie Space Cowboys by Eastwood, He belong to a special group called the, Society of Experimental Test Pilots.

On may 11, 1962 over 49 years ago The late Captain Joseth Albert Joe Walker reported that his job was to confirmed UFOs during his X-15 flights. He also reported and filmed around 6 UFOs when he was breaking his 50 miles flight in April in 1962 of that year.

Retired Captain Eugene A. Cernan He was a highly distinguished officer and received many awards including Induction into the U.S. Space Hall of Fame, the *Challenger* Center's "Salute to the U.S. Space Program amongs many others and was also a member of the, American Astronautical Society, including as well Society of Experimental Test Pilots.

Retired Captain Eugene A. Cernan who was also an Astronaut and Comander of the Apollo 17 mission had this to say when ask about unidentified flying objects Quote I've been asked (about UFOs) and I've said publicly I thought they (UFOs) were somebody else, some other civilization Unquote

I have said on many occasions that its good to be sceptical when ever you are faced with a subject that has to do with UFOs but consider for a moment the following.

We have statements from Astronauts admitting that there is a secrecy that is impose on these. You also got these professionals officers coming out publicly coming out with information that would most likely ignore as either fiction or just they, as Dr. David Morrison did state Quote: Mostly it is people who want to sell you their books or tapes. Also, some people apparently enjoy fooling others. Unquote

Do you accept that these heroes this Officers that had served their Country so loyally And have put their life at risk not only in wars but as Test Pilots would do such a low thing as to lie? And misinformed the public? For what to sell books and videos?

You have also read that these men, this Heroes have received great Honours put to them by their Country for servicing Their Nation and received the highest commendations such as and including, Distinguished Service Cross with oak leaf cluster amongst other awards.

I doubt very much if any of them would betray their ethics in fact after doing my research on those Researchers and Officers I have to conclude that they are telling us the truth of what they saw, reported, and filmed.

Richard C. Hoagland contribution to NASA

I now like to draw your attentions to the Pioneer 10 Spacecraft Mission. This Spacecraft was launched on March 2 1972, it was launched as I recall from an Atlas Centaur rocket and it was launched from Cape Canaveral. So Pioneer 10 became one of NASA greatest achievements not just because it was the first spacecraft to probe the planets of our own solar system but also because of the great amount of data that was sending to Earth and then we started to discovers things about our neighbour planets and our cosmos around us that before we could only guess, and I guess we done a lot even today by the scientific community however much prejudices still exist by the few.

I have place the link here from NASA which I have already acknowledge how ever just so that we don't forget the purpose remember that what NASA Scientist said about Mr. Richard C Hoagland, the he never work for NASA and he knows nothing of science, correct?

There is no question that the Pioneer 10 message board was created by, Dr. Carl Sagan Linda Salzman Sagan and Dr. Frank Drake. However this was a Pioneer 10 project in which the ideas of the whole group employed by NASA that worked on this projects are discuss and put forward I quote the Acknowledgement.

Quote: M. Acknowledgments

"We thank the Pioneer Project Office at Ames Research Center, especially Charles Hall, the Program Manager, and Theodore Webber; and officials at NASA Headquarters, particularly John Naugle, Ishtiaq Rasool, and Henry J. Smith, for supporting a small project involving rather longer time scales than government agencies usually plan for. **The initial suggestion to include some message aboard Pioneer 10 was made by Eric Burgess and Richard Hoagland.** A redrawing of the initial message for engraving was performed by Owen Finstad; the message was engraved by Carl Ray. We are grateful to A. G. W. Cameron for reviewing this message and for suggesting the serifs on the solar system distance indicators, and to J. Berger and J. R. Houck for assistance in computer programming." Page 201 JPL Technical Memorandum 33-584, Vol. I. page 201.**Unquote.** I have provided the link here for you to further research this topic and if you are a student doing a test remember to do a proper research because there are lots of distorted websites out there and they have not been put there by accident.

Technical Memorandum 33-584 Volume I Tracking and Data System Support for thePioneer Project Pioneer 10—Prelaunch Planning Through Second *http://ntrs.nasa.gov/archive/nasa/casi.ntrs.nasa.gov/19730011461_19730011461.pdf*

This is another attempt to distort and discredit Richard C Hoagland and what's worst is that he is downgrading and humiliating Mr. Richard C Hoagland knowing perfectly well has his statements against Mr. Hoagland its just a Hoax.

Here I have provided a website from NASA which confirms that Mr. Richard C Hoagland has contributed immensely to our scientific community.

Before I continue on Mr. Hoagland bio I urged you to look up this site on the Internet and read the very bottom about who the Author is For those of you who don't have an Internet or you are not close to one here is Mr. Richard C Hoagland NASA profiles:

http://www.hq.nasa.gov/alsj/LM22_Moon_M1-15.pdf

Title The Moon

By Author Mr. Richard C Hoagland

**The Author M-15
Richard Hoagland is former staff lecturer and
Curator of Astronomy and Space Science at the Springfield
Museum of Science in Massachusetts.
He subsequently was Assistant Director of the Gengars Science
Center and the Planetarium at Cildren`s Museum, Hartford, Conn.,**

and deviced several major programs to modernize planetariums in the U.S. His innovations include techniques described as "a major breakthrough in the field of planetarium programming and simulation" in the journal Sky and Telescope. A writer and lecturer, Mr. Hoagland is a consultant on astronomy and space science to museums, planetaria, and the aerospace and broadcasting industries.

The other issue is concerning Mr. Richard C Hoagland on whether he and M. Bara was an employee of Nasa

Quote: **is not by former NASA employees** Unquote.

This job of looking for all the data that NASA has it not easy since there is so much written material to cover and assuming that someone has not been tempering with any data regarding former employees who had this should not be an impossible task.

Also the wording of what constitute an Employee is very important as well because especially bureaucrats and politicians are masters of the art in putting out there 20 minutes of talking on a question and not answering to it, just watch your politicians debate and you will see what I mean. This is another source that I found regarding Mr. Richard C Hoagland Check the thread below and read page 17.

VI. GUEST INVESTIGATORS AND VISITORS

http://ntrs.nasa.gov/archive/nasa/casi.ntrs.nasa.gov/19760067279_1976067279.pdf

This particular paragraph refers to Investigators and Visitors and it gives a list Scientific people that came there to pay a visit to the Planetary Research Center in this group of special visitors are very prominent people of our scientific community, there are Doctors, a Professor named William McCrea, Dr. William McKinney, Dr. William Hartman, Dr. Frank Gifford, Dr. Gerald Coupinot, Dr. Josette Hecquet, Dr. Jan Eric Solheim, Dr. Ernest Both, Dr. James Elliot, Mr. Peter Woiseshyn, Dr. Joseth Veverka, But guess what? In this list of very imminent scientists also appears Mr. Richard C Hoagland and this is what it says about this eminent person.

It reads and I quote: Mr. Richard Hoagland (**Who is also a science writer connected with projects at the Goddard Space Flights Center**) Unquote.

The question in my mind is. Why is it so hard to find any references on the projects that Mr. Richard C Hoagland did with NASA? and does this means that Mr. Richard C Hoagland was indeed a NASA employee?

Full time? Part time? Casual? or could he had been contracted?

Richard C Hoagland was the main man responsible for naming Orbiter Enterprise at the time when he was science advisor for the CBS news. This is also stated at the NASA site on page 2.5.2 entitle **The Orbiter Enterprise**.

http://www.hq.nasa.gov/oia/nasaonly/itransition/Shuttle_Historic_Facilities_Roll-up_Report.pdf

But you there has been very little references out there acknowledging Richard C Hoagland in his part in the naming of The Orbiter Enterprise. *http://www.aerospaceweb.org/question/spacecraft/q0288.shtml*

The bottom line is that it was very hard to find references of him and it could be just a simple mistake of just not putting his full name as key words to make it accessible for any one especially the very young and to read what in what project at NASA he was involved it. Richard C Hoagland has contributed much to the scientific community and to the space program including NASA and knows much of science. However since his investigation of the face on mars which have prove to be correct, plus his book Dark Mission has upset many in NASA. It is a fact that Mr. Richard C Hoagland was indeed a former NASA consultant and was indeed involved in various NASA project.

Mr. Mike Bara who's professional Biography includes working for 25 years as an Aerospace Engineer and as well he was also an Engineer Consultant for Boeing and many other companies as well. The point is that he did work as an engineer and like his resume says he was an Aerospace Engineer.

The reasons why NASA is so upset at Mr. Mike Bara was in an interview that he had with a former NASA scientist Dr. Ken Johnston who was also a former Manager of the Data and Photo Control Department at NASA Lunar Receiving Laboratory during the maned Apollo Lunar Program.

Apparently according to Mr. Barra, Dr. Ken Johnston was order by NASA to destroy some photographs which NASA did not want the public to see. In those photos they show some anomalies and ancient structures. As a result of making his findings public Dr. Ken Johnston employment with NASA at the JPL's Solar System Ambassador SSA for short. The bottom line here is that Dr. Ken Johnston did provided evidence of huge structures which was photograph on the moon and he has also claimed that NASA has been very busy airbrushing all the photographs from the moon. One has to wonder, Are there any Ancient cities on the moon? why would NASA feel the need to airbrushed those photos?

Climate Change Disclosure

Every time this subject comes up it reminds me of the false promise that our Prime minister Julia Gillard made just before the 2010 Federal Election, she said "There will be no carbon tax under the government that I lead" since then with the help of the Greens the carbon tax did go through and what I am afraid is that countless jobs and manufacturers will take their business elsewhere and the looser will be the ordinary workers men and woman out there that has to pay their bills and its going to be even more tuff and more pain still to come. Allready there is much talk about electricity bills going up, but just about everything will be affected by the carbon tax con. However there is the question of whether Julia Gillard is in breach of our Australian constitution and if so shouldn't there had Parliament been dissolved and our Prime minister be held accountable? If this is so why doesn't the opposition done something against this? Just my thought. But it makes no sense to me for our Government will close down the Hazelwood coal fired power station which is a perfectly good working power station and yet we in Australia are going to sell thousand of tons of coal to China, were is the sense in that? In other words China will end up having cheap coal power while we in Australia are going to pay electricity through our nose, this is the most comical idea I have ever heard and I ask the question, are our politicians insane? Or is someone else pulling their strings? You know you may not believe this but China right now are buying a lot of farmland here in Australia and that is why China is so powerful because they don't have the nonsense ideas that we have in Australia.

Is Climate Change the result of human activities resulting in the increase of gas emissions? Does this means that humanity is to blamed for the global warming climate change? What about the change occurring in all our planets as well as their moons perhaps the environmentalists conservation lists when starting to look at the media information's showing global warming created by industries and manufacturing they must be thinking something is wrong, something is not right. Consider for the moment that the polar ice caps of Mars are melting and that the ice caps are also withdrawing from its previews positions several miles every year much more quicker than what's its happening to our planet Earth it begs the question, Why are Mars polar caps melting? What is causing the melting of Mars ice caps to retreat so quickly?. It is a fact that in the artic regions of Earth our planet had a tropical climate just 4,000 years ago? that the fossil plants were found inside the stomach of tropical animals. As those found in Siberia in 1901 they were found to contain in their stomachs that the Mammoths had been feeding on tropical plants. So firstly for a body to be preserved in such good condition some sudden impact in our climate change must have occur, suddenly without warning, what was it? Could a giant celestial object such as a comet may had been responsible for such a catastrophic event?

So there are several questions that need to be ask which has not been dealt with by the mainstream scientific community that support the views that human activities are responsible for climate change.

Planet Mercury

Is it possible that the polar caps of today were once hot tropical climates? If so were would those polar regions had been before the polar caps shifted? Dr. Sami Solanski from the Max Planck Institute in Germany has prove that the core samples taken from the Artic and Antarctic regions showing the solar activities higher in eight thousand years meaning that the only time that the sunspots was at its most intense was eleven thousand years ago. Planet Mercury for example had a visitor in 2008 which was messenger spacecraft. In that year NASA issued a report regarding planet Mercury, the magneto sphere including the magnetic field of the planet mercury during the spacecraft messenger it shows to be very different from the mariner 10 spacecraft observations. The spacecraft messenger found great differences in mercury magnetic fields.

Planet Venus

From the SRI International article by Alice Resnick in 1999, shows that planet Venus has had a 2500% increase in Green Glow, it has been increasing from 1978 to 1999 so what does this information mean you might ask. It means that the atmosphere of Venus is changing whether this means that the atmosphere is changing to something that is breathable is another question and remains to be seen because at present the sun has been having huge CME many times cause by huge passing comets which NASA keeps denying at every turn.

Planet Mars

The data and photographs by Don Savage et al, from NASA official Hubble Space Telescope website shows that the polar Caps are melting and also is showing that Planet Mars is growing clouds and ozone. Years ago there were not one single cloud but now the new data that we are receiving the change on Mars is tremendous. In fact the change on Mars is so great that the mainstream media is referring to the weather changes on Mar as Global Warming which is a fact that Mars is warming up. Perhaps NASA can explain to us why are this changes are happening on Mars? Whys is Mars having Climate Change given that Mars has no SUVs. On December 6 2001 Space dot com an article on Mars by Britt Robert Roy "Mars Ski Report: Snow is hard dense and disappearing.

Planet Jupiter

Dr. Philip Marcus in USA Today, reported that planet Jupiter's white oval were disappearing from September 1997 and September of 2000, and then they all melted into one, Dr. Philip Marcus did predict that the Global Warming on Jupiter would cause the planet temperature to climb by 18 degrees in ten years which it did. It would be interesting to ponder what would happen to our world if the temperature here would rise by 18 degrees? More importantly why is it that planets which are so far away such as Jupiter are having climate Change since there are no SUV. There are also massive huge storms in Jupiter in scale much larger that what they once were which are also affecting the colours of Jupiter's rings. So what is causing Global Climate Change in Jupiter? And why are the Storms getting ever larger? And what is causing the colours in Jupiter's ring to change?

Planet Saturn

Dr. Ed Sittler et al at NASA has stated that the Plasma Toras in Saturn has grown by a 1,000% much denser from 1981 to 1993 that was around

19 years ago I wonder how denser it is now, or whether the Plasma Toras has grown even more?

Dr. Jan Uwe Ness from the University of Hamburg in Germany did confirm that in 2004 that the Planet Saturn is emitting massive amount of X Rays that is coming out from its equator.

Planet Uranus

When Voyager 2 orbited around Uranus back in 1986, Dr. Erich Karkoschka from NASA University in Arizona he gave a description of what Uranus look liked from the photographs that were taken by Voayer2, Uranus was "Featureless as a ball"

So why would a scientist describe Uranus in such a fashion? Perhaps NASA did not use the correct wavelength the view Uranus? After all if you look at that photo all you see is a snow like ball with no feature. However photographs taken some thirteen years latter revealed a very different planet especially the planet upper atmosphere. Dr. Erich Karkoschka describe a very different Uranus on photographs taken in 1999. As follows "Hit by huge storms really Big, Big Changes"

So why is it that the planet Uranus is having massive huge storms that was not detectable by Voyager 2? After all Voyager 2 had the advantage of orbiting the entire planet and from different angles, so Voyager2 should had detected weather planet activities.

Planet Neptune

In the year 1989 when Neptune was photograph it showed very little amount of cloud activities in the planet atmosphere. However, this is very interesting, when the planet Neptune was photograph again from 1996 to 1998 including 2002, those photographs shows that Neptune had huge amount of cloud and weather activities, all in all the photographs taken on the infrared wavelength showed a 40% in brightness and instead of scientist admitting the obvious that the Climate Change in Neptune is the result of our sun, instead they wrote "Seasonal changes that are massive but insisted that those changes are not well understood"

Planet Pluto

Pluto used to be a member of our family of planets until it was demoted. In an article written by the BBC News January 27, 2001 entitle, Pluto Dismissed as a Lump of Ice The American Museum of Natural History in New York describing Planet Pluto "Pluto has more in common with comets than planets as it is relatively small and made of ICE" I assume that decision was taken by someone with a PhD in astrophysics so this then bring up a whole series of different issues which those that are making this decision should answer. So I like to pose a question or two, First I like to know why is it that those orthodox scientists still hold to the notion that comets are made of dirty snowball? Since the data that came out of DEEP IMPACT clearly proves that Tempel 1 has hardly any water at all to produce the coma and Tempel1 is just a hard rocky object. But since they reffere to planet Pluto as a comet, how is it that the data that Dr, James Elliot, MIT/NASA shows the following. Pluto is experiencing Global Warming and its atmosphere pressure has increase by 300%. So with that sort of data and given that those scientists still believe in the myth that comets are ice, shouldn't Pluto be forming a coma? And be developing the hugetail as the iced body begins to heat up? Also why is Pluto heating up, being that Pluto is so far away.

ROBBING PLUTO OF ITS PLANETHOOD

Melbourne's most severe weather ever seen

On the night of December 24 of this year 2011, there was continued thunder storms with the loudest thunder I have ever heard in Melbourne. With heavy rain and the thunderstorms continue well into the afternoon and during our Christmas meal on December 25, that we were having in our parents home at the time. At 15:56 the thunder became extremely hostile and with the massive thunderstorm came huge hail stones the size of which were very large the size of which we never seen before in Melbourne.

The storm was such that it dented the cars park outside, on both the top, back and sides of our cars luckily the damaged was very light to our vehicle as compare to other people which had its windows smash by the huge hail stones.

At 6:45 PM the rain continued to fall extremely heavy once again with much thunder and lightening the day became night then at 7:15 PM the sun suddenly broke through the clouds like a ray of hope, but the thunderstorm continued, Indeed we have never experience a weather like this one during our Christmas season. I do remember way back in the seventies when the temperature would be in the midst thirty degrees plus, But during the eighties and even as recently as on the 2010 the temperature was very cool in the 20 degrees Celsius but all that changed in this Christmas eve of 2011.

Later on more bad news was to follow, Planes had been grounded with massive flooding reported and a large amount of vehicles was reported damaged, and hundreds of calls for help were received by the SES. Later on more bad news arrived as the weather bureau reported that huge hail stones the sizes of tennis ball had cause service damaged across Melbourne including mini tornados were accompanying the huge thunderstorm that hit many parts of Melbourne including North West and through to the North Eastern suburbs had seen the worst of it. Lots of roads were also reported under water including Bundoora and parts of Plenty road were under water, including also Mount Alexander road which is near Essendon Station were also flooded.

By the days end the SES had received more that 1200 calls for help. This storm I believe was just as bad as the one that hit on March of 2010 which cause havoc and much damaged across the whole of Melbourne including the Southern Cross station that was closed down due to thunderstorm and hail stones damaged.

By the end of the day the SES received 3,000 reports for help, and the SES drafted from 250 to 300 SES volunteers from all over Victoria, many thanks to those men and woman that have volunteer their services to help all Victorians, they do a remarkable job during very difficult circumstances.

Its also been estimated that the damaged bill is going to cost somewhere into the tens of millions of dollars

Geo-engineering May 2009 by the Late Mr. Zecharia Sitchin Global Warming

A new term, "Geo-engineering," has come to the fore at various international meetings dealing with 'Global warming' (now more correctly addressed as Climate Change).

The term was officially embraced by the new U.S. presidential science advisor, John Holdern. Speaking at a recent international conference in Bonn, Germany, he revealed that "Geo-engineering" is among the "extreme options" under discussion by the U.S. government: Using space-age technology yet to be devised, he said, *"particles will be shot into the Earth's upper atmosphere* to create a shield that will reflect away from Earth the Sun's warming rays." While such extraordinary measures would be only a last resort, "we don't have the luxury of taking any approach off the table," Dr. Holdren said.

Such new space-age ideas *duplicate technologies that have already been used 450,000 years ago!*

Back to the Anunnaki

The audacious idea of protecting a planet thermally by creating a shield of particles in its upper atmosphere is not as revolutionary as it seems. It was, I wrote in my 1976 book *The Twelfth Planet*, exactly the reason why the *Anunnaki*—"Those who from Heaven to Earth came"—had come here some 450,000 year ago from their planet Nibiru.

On Nibiru—'Planet X' of our Solar System—the problem was the opposite one: Loss of internally generatedheat due to a dwindling atmosphere, brought about by natural causes and nuclear wars. Nibiru's scientists, I wrote, concluded that the only way to save life on their planet was to create a shield of *gold particles* in their upper atmosphere. It was in search of the needed gold that the "gods" of the ancient peoples had come to Earth. Basing my conclusions on Sumerian and other texts from the ancient Near East, I wrote that the Anunnaki began to arrive on Earth some 445,000 years ago, establishing settlements in the E.Din (later Mesopotamia) and mining gold in southeast Africa.

As I have written in subsequent books, "modern science is only catching up with ancient knowledge." The idea of 'geo-engineering' is borrowed from technologies of the Anunnaki.

© May 2009 **Zecharia Sitchin**

Reprinting permitted on condition that
© **Z. Sitchin** *is indicated.*

Megaliths Structures, Who Built Them?

I you look up the meaning of the word Megalith in any of the dictionaries it will tell you that megalith is just a large stone that was built around the second millennium B.C. and that may be all the information that you may have and if you ask your teacher he/she would probable tell you the story of how they might have been built based on what he/she has been told by their peers. In fact you don't need to have a PhD to know that those megalithic structures are found around the World at a time when people are said to be very primitive and that's the view of our scientific community is and what we have been taught in our schools and universities even to this day. In fact those massive stones that weight as much as 2,000 tons, scientists with PHD after PHD are still teaching and writing books, that those massive structures were built by primitive people using fibber ropes and stone stools. Its like Alice in wonderland that such a primitive people could have ever done such an enormous work. Even to lift such a massive stone you just cannot do it using ropes much less cut a stone to such precision. The one massive structure that comes to my mind is the unfinished Obelisk found in Egypt. Its just massive it weight at 1170 tons and it has a crack on the front but and it was just left there. On this Obelisk all the four sides are perfect but its pretty obvious when looking at this unusable structure that stone tools could never had done such work. This Obelisk is 15 feet deep and yet Egyptologist tells us that this Obelisk was built using stone tools so I like them to answer, how would they lift this structure ? more importantly is, how would you cut this massive stone from its base? It is just even in today's technological advancement we have no such tools as to cut the base of such a structure as this Obelisk, so how is it that Egyptologist still maintain that this structure was cut, lifted, and transported using primitive tools. The of course there is the much talked about, the Pyramids at Giza, according to the orthodox views of the Egyptologist the three Pyramids at Giza built by Khufu builder of the first pyramid, Khafre builder of the second Pyramid and Menkaura builder of the third Pyramids. Now the Egyptologist tell us that all three Pyramids were built within one hundred years which is the life time of all three of these Pharaohs and it took six million massive stones to make those Pyramids, so lets work it out.

100 years X 365 days in a year = 36,500 days
36500 days X 24 hours in a day = 876,000 hours
876,000 hours X 60 seconds in one minute = 52,560,000 seconds

52,560,000 seconds ÷ 6,000,000 stones = 8 minutes 76 seconds.

So to construct those massively huge megalithic Pyramids structures you will need to cut, transport, and put into one stone every 8 minutes and 76 seconds day and every night for the hundred years.

It is very obvious to me that based on this numbers that primitive people could not have don't such an enormous undertaking that we in our high technological advancement could not do. There has had to be an advance civilization prier to our own and we made had found it as the photographs from the missing lunar tapes confirm.

What does the Government knows about Missing Lunar Orbiter Tapes?

48,000 lbs or 21772.43376 kilograms Missing Lunar Orbiter inside an abandoned McDonalds restaurant near Moffett Federal Airfield?

Were those massive megalithic structures build by Aliens?

Is Comet lovejoy just a little snowball if so how could had survive such massive temperature?

What's causing Climate Change in our Solar System?

Are the theories on comet size wrong?

Does the Oort Cloud even exists? Or is this theory just another myth?

I am sure that in time we will find the answers to this and other mysteries in our endeavour to seek the truth.
An emotional plague afflicts people whose belief systems are so rigid they ignore relevant facts and become enraged if anyone challenges their beliefs.

Wilhelm Reich

www.ingramcontent.com/pod-product-compliance
Lightning Source LLC
Chambersburg PA
CBHW021009180526
45163CB00005B/1941